COMPUTER SCIENCE, TECHNOLOGY AND APPLICATIONS

ARTIFICIAL INTELLIGENCE ENGINEERING FOR POSTSECONDARY EDUCATION DIGITAL TRANSFORMATION

COMPUTER SCIENCE, TECHNOLOGY AND APPLICATIONS

Additional books and e-books in this series can be found on Nova's website under the Series tab.

COMPUTER SCIENCE, TECHNOLOGY AND APPLICATIONS

ARTIFICIAL INTELLIGENCE ENGINEERING FOR POSTSECONDARY EDUCATION DIGITAL TRANSFORMATION

JESÚS ALFONSO PÉREZ GAMA
LUIS GIOVANNI ROZO PARDO
AND
LUIS CARLOS GUTIÉRREZ MARTÍNEZ

nova science publishers
New York

Additional color graphics may be available in the e-book version of this book.

Library of Congress Cataloging-in-Publication Data

Names: Pérez Gama, Jesús Alfonso, author. | Rozo Pardo, Luis Giovanni author. | Gutiérrez Martínez, Luis Carlos, author. | Nova Science Publishers.
Title: Artificial intelligence engineering for postsecondary education digital transformation / Jesús Alfonso Pérez Gama, Luis Giovanni Rozo Pardo, and Luis Carlos Gutiérrez Martínez.
Other titles: Computer science, technology and applications.
Description: New York : Nova Science Publishers, Inc., [2019] | Series: Computer science, technology and applications | Includes bibliographical references and index.
Identifiers: LCCN 2019042868 (print) | LCCN 2019042869 (ebook) | ISBN 9781536163902 (Hardcover) | ISBN 9781536163919 (Adobe PDF)
Subjects: LCSH: Artificial intelligence--Educational applications.
Classification: LCC LB1028.43 .P46 2019 (print) | LCC LB1028.43 (ebook) | DDC 378.0285--dc23
LC record available at https://lccn.loc.gov/2019042868
LC ebook record available at https://lccn.loc.gov/2019042869

Published by Nova Science Publishers, Inc. † New York

In Memoriam:

My Parents José de la Cruz Pérez Durán and Emilia Gama Niño.

This book is dedicated to my children Hada Jessica, Byron Alfonso, and Osmar Alberto. Also, to my beloved £eidyvina Verdugo.

To our Fellows from the "Fundación de Educación Superior San José": to Lecturers and Officers; Principal Francisco Pareja, Chancellor Carlos F. Pareja; specially, to San José EIDOS Research Team, including: Claudia Lucía Caro, Guillermo Hoyos Gómez, Hernán Remolina Caro (APHROS), Manuel Neira Aponte (BI Planning), Victor Yockteng and Yicel Rodríguez. To all of them, a debt of gratitude. To July Andrea Espinosa-Sánchez, our Secretary, for all her permanent help.

Jesus Alfonso Pérez Gama

To my Parents, Luis Jesús and Raquel, always a support.

Luis Giovanni Rozo Pardo

To my Parents José Orlando Gutiérrez and María Yubisa Martínez.

This book is dedicated to my wife Eloira Cogollo Caro, to my daughter Isabella Gutiérrez Cogollo, whose understanding, support, and love have been vital in my process as an entrepreneur, academic and researcher throughout my life.

To our leaders of the "Foundation of Higher Education of San José": to professors and officials; The director Francisco Pareja, the chancellor Carlos F. Pareja; In particular, to the EIDOS research team in San José.

To my partner of great initiatives by the academic communities in Latin America, to Old Willy Marroquin, "Power for the people."

Luis Carlos Gutiérrez Martinez

CONTENTS

List of Figures

LIST OF TABLES

FOREWORD A

Francisco Alfonso Pareja González

The higher education institutions, the government and mainly the managers of educational processes are facing the new world scenario leading by the ICT, with the fast and challenging of the knowledge society. If they want to be at the forefront, with great topics such as the Digital Transformation (DT), (which is fostered by the rise of an eminent IV Industrial Revolution) they have to adapt, modernize and innovate all the academic, organizational and financial systems to be with effective and efficient responses to the postsecondary education propaedeutic cycles challenges.

The Fundación de Educación Superior San José, is an Institution of Higher Education with more than 34 years, working for the formation of excellent professionals: with more than 20,000 graduates to date, who have contributed to the development of this country from their professional skills and social ethics with our sanjoseísta seal, under the propaedeutic cycles strategy: they are magnificent results. They constitutes our essence and reason of being; the educational proposal is based on the implementation of pedagogical strategies with updated models, which are based on modern learning issues, participation, respect and good relations in the diversity of thought or initiative, approaching to scientific, technical and technological knowledge and the development of a dialectical relationship, with the social ecosystem environment: economic, industrial and political that are benefit the most inhabitants, at local, regional, national wide.

Our Institution has been appointed as the first HEI –T and T in Research, according to the ranking of the S & T Observatory, and also with several awards, distinctions and national and international recognitions.

Aware of our socio-political responsibility with countries and nations, we have recognized the need for participating pro-actively in the processes of the real transformation (social, economical and technical) regarding the welfare of our students, their essence as individuals and their formation with ethical, scientific, technological and social criteria. They are able to make their own decisions, taking as a starting point a modern culture of teaching and learning that strengthen their personal and professional lives and acting as protagonist agents of a new society.

We want with the contribution of the book *"Artificial Intelligence Engineering for Postsecondary Education Digital Transformation"*, assume on the part of the leaders of educational processes, a position of conscious for changing new educational models that must be considered, a new type of institutions, of major level, which be considered more competitive, inclusive, open, diversified, and regionalized. Our perspective, is the internationalization (according with our Social Projection model) in order to have more influence in the formation of new citizens of the knowledge society in the global one, concentrating all efforts in the redefinition, reconceptualization and transformation of postsecondary education model, under the principle of high quality, with a clear mission and social educational project that develops thinking and feeling of all the members of the living communities in the current world scenario.

This book is presented as a dynamic that from the perspective of the DT, the identification, evaluation, modernization of the needs and at the same time, seeks to develop towards a system that is at the forefront, and allows to lead processes of quality, productivity and competitiveness and that also has the added value of giving a closer vision to the remote regions, away from science and technology and with greater needs, to propose solutions from DT to the real needs of our society, which looks so unequal and that we suffer today, allowing access to the PSE to those people, who for various situations or factors (such as socioeconomic, cultural or geographical) who historically have not been able to project a more dignified quality of life.

Finally, it is important to recognize and to thank the team of authors who have worked with high rigor for several years in research processes around the DT to strengthen contact with the globalized world, so that the Institutions become a bridge for new views of the system education that were previously marginalized (i.e., digital gap), and now can use one of the available means to democratize the all education level and reduce the gap: digital divide. The research of Fundación de Educación Superior San José. has focused on the most complex and complicated problems of the Higher Education, such as student desertion and permanence, low quality and low competitiveness, applying mathematical and computational models with multilevel Architectures, Knowledge Engineering and Artificial Intelligence: building new systems, methodology, tools and instruments which have been implemented in several universities in this country.

The IV industrial revolution brings with it a technological and computerized reach superior to the level of current intelligent machines and systems, nowadays elements such as genetic sequencing, nanotechnology, sustainable development, renewable energies and other activities that are becoming more important are involved. For the human being, therefore, it is recognized that the classification and the indicators of digital maturity imply reference points in time, for comparison purposes, which gives up an idea of the progress in DT, the academic productivity of the teachers, the students and the institutions, along with competitiveness.

Based on the above, several metrics are generated to determine the category and the desired level in the institutions, so that the expert evaluators will express their perspectives in terms of complexity, cost, time, income expectations and risks, categorized in High development, Good development, Poor development, when be the opportune moment for evaluating purposes.

Finally, the importance and the plus of the use of current extreme technologies is recognized, offering students the opportunity to be better educated and trained professionals to lead this transformation in the advent of the IV Industrial Revolution.

Some of the leading fields to promote new knowledge discovery and development in DT are: knowledge and artificial and computational Intelligence engineering, mobile communications, large-scale data analysis (Big Data), advance social platforms, advanced robotics, machine learning, IOT, Learning analytics and augmented reality.

This book is our great bet, for the reflection, transformation, development and consolidation of the Higher Education Institutions, with the aim of contributing to the construction of peace, justice and equity for a society that is more tolerant and respectful of the human rights and of the environment and with social responsibility of our globalized world, the publication of this book is considered an important contribution of the Fundación de Educación Superior San José towards the transformation of Colombian Higher Education Institutions and the rest of the world. We have implemented the full DT model inside FESSANJOSE, as our bet to the first Colombian Digitized University, transformed to be of major level Institution.

Biographical Sketch

Francisco Alfonso Pareja González

The Congressman is graduated as a Mechanical Engineer, Universidad de los Andes at Bogotá. He has a Specialization in Adult Education. Also he has acted an Educator with more than 30 years of experience at all educational levels starting from the SENA (National Apprenticeship

Service); He has performed managerially an Educational entrepreneur, He has founded several institutions: The Triangle Institute, The San José Higher Education Foundation, also the Model Comprehensive School for Young People and Adults, He lead the Springfield and acting as Language Center CEO, which is a Company R&D+I, Ibest, Also he is a Researcher and Social Leader on Educational issues and the related social impact.

Large charges exercised by popular election: Councilman of Bogotá, in 2 consecutive periods from 1995 to 2000, He was also a Congressman to the House of Representatives for Bogota from 2002 to 2006.

FOREWORD B

Manuel Martínez Niño
CINTEL - Executive Director, Bogotá, Colombia

In my opinion today, one of the great challenges that countries, governments, companies, academia, and ordinary citizens have is: to achieve the effective inclusion of the society in what we know today as *digital transformation* (DT). It is fully demonstrated that the proper use of *information and communication technologies* (ICT) in all the productive sectors of a nation necessarily leads to the improvement of the levels of quality, competitiveness, and productivity of the same.

That is a complex challenge for emerging countries, given the existing socio-economic difficulties of access, use, training, and availability of ICT; the efforts that countries such as Colombia are carrying out in the search for the effective development of a *digital economy* are great; it is a task that will take time, especially since the DT has as starting point the transformation of the person into a *digital citizen*.

The academic world has an advantage in this sense. The students, mostly young people (i.e., the millenniums), are already digital natives; since their early years they have had the opportunity to access, interact and play with the digital world; this is a fundamental aspect that allows a more rapid migration to the DT given that students have experienced and know the great potential of competitiveness that DT offers them for all their actions.

The universe immersed in DT contemplates a wide range of emerging digital platforms with great capacity to generate disruption: *artificial intelligence*, *cloud* computing, the Internet of things and people, *big data*, *data science*, *analytics*, social networks, ... etc. *Artificial intelligence* is part of the DT, and is undoubtedly a field of research and technological development since the last century, that is already demonstrating a strong impact on society: the gradual displacement of work traditionally done by humans now by robots, is every day more latent; the accelerated technological development is unstoppable and it is necessary. Our challenge is to know how to assimilate it in an intelligent way.

As a professional in *systems engineering* for about 40 years and promoter of the *digital transformation* in this country, I am honored and pleased to write this short prologue for the book that Engineer Jesús Alfonso Pérez Gama, professor of mine at the time of student at the District University Francisco José de Caldas in Bogotá, has written and in which he demonstrates and focuses his great experience towards the benefit of *higher education*. Professor Pérez Gama recently received the highest distinction from the Colombian Government through COLCIENCIAS: as Emeritus Researcher. This book is a testimony at more than half a century, of his academic work with rigor and scientific discipline, part of the results and efforts of research and innovation. Many publications defended in national and international events endorse his tireless and continued contribution in the academic world.

Biographical Sketch

Manuel Martínez Niño

Systems Engineer of the Francisco José de Caldas District University (1977), with studies in Telecommunications Specialization with emphasis on Distributed Processing Systems at the Fridericiana Technical University of Karlsruhe, Germany. Scholarship by the German Foundation "Friedrich Ebert Stiftung" (1980-1983). He has extensive managerial, technical, and marketing and sales experience in the ICT sector, obtained after more than 40 years of working and managing multinational and national companies in the sector. Among other organizations, in the private sector he has been linked with General Motors Colmotores, NCR of Colombia, Wandel & Goltermann, Comware, AT&T, and Sun Microsystems. In the Colombian public sector, he has worked with the Ministry of National Defense and with the Ministry of Finance and Public Credit.

Since October of the year 2000 Dr. Martínez Niño is the Executive Director of the Center of Research and Development in ICT, CINTEL, in charge of identifying, leading and consolidating the lines of action of the Center with which this corporation, that brings together Universities and public and private companies of the sector, can support and boost the growth of the ICT industry in Colombia.

CINTEL chaired by Dr. Martínez Niño, has developed more than 500 innovative ICT projects nationwide, as well as the realization for 17 years, of the most important International ICT Congress in Colombia and the Region, with great international recognition, ANDICOM.

In order to know and assimilate global experiences related to the use of ICT as social and economic development-generating tools, he has led multiple missions in countries such as Sweden, France, Germany, Spain, Italy, Canada, the United States, Brazil, Mexico, South Korea, China, and Japan, among others.

He successfully has completed the high-level training program in Science, Technology and Innovation Management sponsored by COLCIENCIAS in the year 2006, in partnership with the universities El Rosario and de Antioquia of Colombia, the University of Sao Paulo in Brazil, and the universities of UCSD and Harvard in the U.S.

He has received numerous awards and recognitions: in January of 2010 he was awarded by the Ministry of ICT, the Medal to the Merit of the Communications Manuel Murillo Toro First Class Gold Category, in recognition for his contribution to the development of the ICT sector in Colombia.

As Executive Director of CINTEL within the framework of ANDICOM 2016, he received from the Senate of Colombia the award "Order of the Congress of Colombia in the rank of Commendator".

INTRODUCTION: DIGITAL TRANSFORMATION INSIDE UNIVERSITIES

ABSTRACT

Our Proposal in the Book is to (1) present, initially, conceptual foundations for artificial intelligence, expert systems, and knowledge engineering and management, and discuss about high quality in education; then, (2) discuss the basics of our vision and perspective about *higher education* (HE): the battle for the future with *digital transformation* (DT); next, (3) present our central Chapter 3 on DT: dealing with our dual model of knowledge and data, as befits a HE institution. Below, we present a succinct outline of our architectural model of university DT; this is a dual DT model: curricular digitalization plus institutional digitalization, i.e., knowledge and data. The pillars of the architecture are: *funding, research, entrepreneurship, digital talent* and *social projection (Chapter 7);* recognizing from the start that *knowledge* has its ethos in the university; these pillars correspond to:

- Productive ecosystem of the transformation
- DT that enhance knowledge and innovation in the universities for the habilitation of their digital capacities
- A new economy that requires transformation, also supported on entrepreneurship
- New DT human talent required by the new knowledge and intelligence industry, in which the student hyper-personalization by competences and skills is required.

DT Basics: Ten Views of Our DT Model

Our perspective of the **Digital Transformation (DT):** DT is a complex process that takes place in long ways, and is required for the successful survival of an organization immersed into the IV Industrial Revolution, with the final purpose of being very competitive, productive, and of high quality (HQ). It is the way to the 4th Revolution. It implies the use of ICT[1] to profoundly improve the performance of an enterprise in terms of productivity, competitiveness, and digital changes in all aspects of the day-to-day of the human life.

DT Model includes several views, namely:

1. The Ecosystem of the DT.
2. The structural vision, or the pillars of the university DT: *research, entrepreneurship, digital talent, budget,* and, especially, *social projection of the institution*.
3. Sustainability.
4. The strategic map of the DT, showing scenarios, actors, and vision-mission.
5. The architectural components for DT: eight architectures developed and implemented for the DT Model, applying some intelligent constructs that we have developed and documented in the last 10 years at FESSANJOSE[2], leading DT in *postsecondary education* (PSEd) in our country.
6. Digital 360° architecture of DT Academy and Administration *LOCUS*: This architecture is the *digital portfolio* that implies the organization of the subsystems to obtain better and/or new functionalities based on knowledge to obtain an intelligent behavior. The multilayer-architecture approach is a system of systems (SoS)

[1] Convergence of information technologies (cognimatics, computing and telecommunications, particularly the Internet and computer networks) with their technological base and information or "content", that is acquired, stored, handled, transported and transmitted through these technologies. The technological basis includes hardware and software components, as nanotechnology, microelectronics, computing and telecommunications.

[2] Fundación de Educación Superior San José, U. San José, Bogotá, http://usanjose.edu.co/.

one, which ensures compliance with government policies, norms and standards, in a highly complex social institution with digital assets; this approach describes the subsystems at a higher level, where a system is composed, and with the protocols by which the subsystems communicate. It provides a 360° business vision map and a planning framework for commercial and technological changes.

7. The computational-mathematical perspective of DT, identifying endogenous and exogenous variables and their interrelations.
8. The synthesis, the *Matrix of End-Means* (EMM) that summarizes in DT: 'Where the HE is?' and 'Where can the HE go?'
9. The MIR Matrix: DT Objectives, *impacts-indicators and results* to be obtained with the DT.
10. The dynamic model of the DT system, based on computational intelligence, representing the control system of all the components to achieve the completion of the DT. It is an *intelligent management information system* (iMIS) for PSEd that shows the dynamics of DT control, integrating several multilevel system hybrid architectures, as a space to respond to the solution of HE problems, tending to the HE institutions (HEI) *competitiveness*, specifically pointing out the way these modern technologies can be included for their adaptation and evolution in PSEd in post-modernity, making *governability*, and *teaching* and *student productivity* compatible with educational high quality, the purpose of DT.

The interface *Results of the iMIS* includes: *high quality metrics*, *indicators*, and *values of management*, *desertion*, *answers*, and *plans*. These results can be decisions, information (e.g., about permanence), required knowledge, solutions, inferred plans, policies, and objects of continuous improvement for the sustainability of the transformation and quality, among others.

The *Input Interface* includes data, information and knowledge acquisition, where the attributes, parameters, values and problems (e.g., of desertion) are registered in such a way that they are actively and dynamically obtained and structured. *The problem* is expressed as the type of a decision to be made, e.g., the need for information in the complexity of the *student dropout*: i.e., for students trained in mathematical competences, a prospecting of system variables, as decrease rate in the *excellence - current state* gap.

MEMBERSHIP OF COLOMBIA TO THE OECD: THE ORGANIZATION FOR ECONOMIC CO-OPERATION AND DEVELOPMENT

At first, given that Colombia has many lags in HE, poverty, inequality and income, will join the club of the rich countries? That was an almost impossible and very demanding path.

The *advantages* of OECD membership, mainly, are:

1. International visibility: improvement of the image of Colombia.
2. Many opportunities in terms of investment, employment, economics and socio-technical progress.
3. Better policies for a productive and competitive industry, a better life and in *peace*.
4. Greater confidence and *international influence*: investment funds and access to new sources of financing.
5. More employment and greater economic activities.
6. Important advances in the reduction of poverty and inequality.
7. Access of the country to 23 committees that the OECD has; Colombia implemented 51 public policy measures (laws (anti-bribery law), decrees, resolutions, and CONPES (National Council of Economic and Social Policy) documents).
8. Reception of quality information and excellent socio-economic studies.
9. From passive receiver of policies to contributor: towards its creation.

10. Continuous and coherent improvement: Colombia will be evaluated with international standards and with proven and efficient review mechanisms to evaluate public policies.
11. It makes it possible to honor the new obligations.
12. Help achieve a *peaceful and equitable country*.
13. Reduce *social exclusion* and reduce the *digital divide*.
14. *Recommends* reintegrating the *HE*, broken apart by the Law of Tertiary Education[3].

Colombia Full member no. 37 remains as a success story. Officially it was this May 25, 2018. Colombia will learn on the basis of example of the advanced countries, achieving better perspectives and visualizing better public policies, towards productivity and competitiveness, in line with the OECD which will undoubtedly strengthen the DT in Colombia.

Finally, if Colombia does things well, implementing best practices will accelerate socio-economic and scientific development, with well-being. Likewise, the Government designed to the National Planning Department as its *interlocutor*.

Colombia must make substantial improvements in *coherence* in *productive transformation* and the cohesion of the *MinTIC*[4], *education and industry policies*, with *state efficiency in productivity and competitiveness*.

In conclusion, the OECD is the best intermediary of universal size with influence in the technical reference frameworks: it is the bridge between Europe and Latin America.

THE DT FINAL PURPOSES

The architecture of DT seeks to:

- Extend the benefits of digitization to the entire organization to achieve better competitiveness, and academic and administrative

[3] www.Mineducacion.gov.co.
[4] Ministry of Information and Communication Technologies, Colombia, www.mintic.gov.co.

productivity, led by academic and administrative strategies of transformation towards *high institutional quality*.

- Digitize all the curricula in terms of the competences-objectives of the curricula, measuring their impact on the objective proficiencies of training and determining the added value of the universal knowledge of each subject with their academic credits.
- Do dual self-evaluation (Chapter 5): based on the initial model provided by the NCA[5], contributing to the quantification of each one of the factors and variables involved that makes the *evaluation analysis* possible, towards the HE high quality. Includes the self-evaluation for an institution's products, as one of the migration and continuous improvement instruments, towards a high educational quality that allows defining strategies to reduce operating costs.
- Develop an intelligent reengineering of processes towards their robotization.
- Design and implement smart models focused on the student, in order to improve student's cognitive and academic productivity, and also to reduce costs. Likewise, for the teachers, to improve their productivity in teaching and in the development of knowledge verification methods.
- Reduce the human talent gap, a condition required by DT and the IV Industrial Revolution.

The Chapter 4 is dedicated to the *Portfolio Analysis* of products, PSEd's intermediate and final results in front of *institutional policies*, projects, and *lines of investigation*, quantifying their relative contribution, and the added value by each one of the items. For each case, the respective *relevance and validity* is quantified.

The following Chapter 6 deals with *classification of the maturity of institutions* in relation to DT, taking into account the possibility of evaluating and quantifying the multiple determinations of the DT as a system: this corresponds to our DT 10 mentioned views, making abstractions

---

[5] NCA: National Council for HE Accreditation (Consejo Nacional de Acreditación), Colombia.

from the DT EcoSystem and from *selected scenarios*. It was decided to consider 3 levels of maturity, considering also that characterizations of companies were made to this effect, by reviewing numerous surveys and considerations of prepared studies. Our construct for DT classification is carried out by *experts* who express their insights through fuzzy values.

WHY DT ARCHITECTURE?

Enterprise Architecture (EA): Architecting implies the organization of subsystems or components to obtain better and/or new functionalities. If any component is knowledge-based, an intelligent behavior is obtained. The architecture multilayer approach is a system of systems (SoS) one, that ensures compliance with government policies, rules and standards, in a highly complex social institution with intellectual assets and knowledge processes, which is a usual situation in a university institution; this approach describes the subsystems at a higher level, where a system is made-up, and with the protocols by which they communicate.

EA and *business architecture* constitute a conceptual tool that helps organizations understand their own structures and the approaches by which they work. It provides a business 360° vision map and planning frame for business and technological changes. EA is presented like a system overall configuration of subsystems or components organized in layers, where each one describes an ordered congregation of structures and common functionalities grouped by a purposeful criterion inside the business.

On the way to the new era to be provided by the IV Industrial Revolution companies of all kinds are focusing on the massive and optimal use of ICT in different organizational plans including human talent, organizational structures, processes, inputs, products, services and obviously the business model. Nowadays, most of the top companies have robust DT teams, and close to a half of them are being driven by high ICT and *artificial and computational intelligence*. On the other hand, very few Colombian HEI have tackled the problem of HE competitiveness and productivity, and much less of them have made the transition to the IV Industrial Revolution by taking the leap, thus, they are receiving a tremendous blow due to academic obsolescence. The universities should play an important role in the reduction of the digital divide, and even more, they must design new programs that face the formation of the digital talent required by this new paradigm; or we will be late again to change.

In the last 3 meetings of ANDICOM (ICT International Congress in Latin America[6]), numerous national cases of domestic DT were presented from companies, but, unfortunately, not a sole case on HE. Will the case of computerization in Colombia be replicated here, process which was left at the discretion of transnational computer elites? The PSEd DT implies necessarily the cognimatics (the informatics for the *Knowledge Society (KS)*) for all the companies, and their social insertion in this country.

During several years we have visited hundreds of HEI in Colombia for evaluation of Minimum Conditions of Quality and High Quality Accreditation, which provided us a great amount of knowledge about the academic and institutional management, that we are trying to apply for the development of our DT Model presented here.

Our ***Dual Model of University Digital Transformation, Knowledge-Data***, has the following *value proposal*:

> Post-secondary education of high quality with new pedagogical models, to compete in the IV Industrial Revolution with a valid educational model based on international competencies to be cultivated, duly personalized, in the student.

What we are witnessing today is a seismic change in the global educational and commercial models. Traditional models, including curricula, are being completely discontinued. *Visionary institutions* can see emerging opportunities and transform themselves to exploit these prospects.

Finally, most of our tools have metrics as a quantitative measurement of an architecture, system, component, or process, to determine the degree to which they possess a given attribute. These metrics are one of the multiple

---

[6] https://andicom.co/en.

determinations of a problem. Cognimetrics are often used to measure knowledge, quality and process improvement.

First Manifesto on Digital Transformation in Colombia

For the first time the Colombian industry presented a Manifesto on Digital Transformation in Colombia, during the closing day of the 31st. ICT-ANDICOM Congress 2016[7]. It is a statement of the purposes and needs of companies to continue their digital transformation and, on the one hand, of the challenge of taking advantage of all the benefits from ICTs and their extensions, and the threat this poses to the traditional models of companies and educational institutions that can disappear, as has been happening in the world and in Colombia.

The structuring of relationship models and digital attention with clients and citizens is highlighted, and equally, the adaptation of processes, resources and infrastructure to comply with the new models and the strengthening of the digital ecosystem through an intersectoral platform. As a result of this dialogue, the Manifesto identified 9 basic *actions* to promote DT. These actions are undoubtedly a road map to take Colombia a few steps further in terms of productivity and competitiveness, generating greater socio-economic and scientific-technical development, and an invitation to 'collaborate' more among all for an end and common interest: 'a better country' affirms the Manifesto.

These actions are:

1. *The Digital Individual,* endowed with the competences that must be developed with essential educational and business actions, is the center of the DT.

---

[7] ANDICOM 2016 ICT International Meeting Cartagena Colombia- Center for Research and Development of Information and Communication Technologies (Cintel) chaired by Dr. Manuel Martínez.

2. *The State as an Example Of Dt*, with significant actions with e-government that require developments of state processes towards the citizen.
3. *A More Symbiotic ICT Industry* between supply and demand.
4. *Digital from the Board of Directors:* the intervention of all the organizational levels involved.
5. *Digital Specialists* that should be cultivated since the digital talent gap is very large.
6. *The State as a Promoter* with tax conditions that favor the investment in DT.
7. *Listed Infrastructure* with conditions that enable scientific-technical reconversion.
8. *Catalyst Regulation,* since regulations are needed in the digital framework.
9. *Long-Term Vision* given that the DT is a long day.

These actions can be grouped into the following three main fronts or pillars:

- *Digital relations*: structuring of relationship models and digital attention with customers and citizens.
- *Being digital*: those who transform successfully adapt their processes, resources and infrastructure to support these digital relationship and attention models. Change of culture and approach towards new business models.
- *Strengthening the digital ecosystem*: an enabling cross-sectoral platform, by digitally transforming the government, productive and social sectors, will have better tools to take advantage of the great opportunities of the IV Industrial Revolution, and citizens will have better welfare conditions by closing gaps in inequality and being integrated into the digital world.

The above is magnified with the entry of Colombia to the OECD, as presented in the previous Section 1.2.

OUR BOOK'S STRATEGIES

This book uses extensively the approach to AI engineering that uses mathematical models to try to engineer the AI by quantifying the different constructs, with a clear orientation also by computational models, which allows us to transfer different worlds of a problem, from the abstract to the virtual, and to the real visions. We follow the life cycle mega-stages: *requirements engineering & analysis*, *functionalities* (i.e., behaviors), *architecture design & synthesis*, and finally *VVTL* (validation, verification, testing and legitimization).

CONCLUSION

We have presented in this chapter the preliminaries of the DT: 9 views of the complex architecture, with the DT objectives and macro purposes, the *raison d'être* of the architecture, and the strategies followed for this book.

Chapter 1

BASIC AND CONCEPTUAL ASPECTS FOR DIGITAL TRANSFORMATION

ABSTRACT

This chapter has been divided in three items:

Part 1. Intelligent systems knowledge base, IKBS
Part 2. Basic and conceptual aspects for digital transformation
Part 3. Organizational culture 4.0.

The first is referring to the concepts and elements of *artificial intelligence, expert systems and knowledge engineering and management.*

The second one refers to the indispensable elements for the DT.

And the final, refers to the aspect of organizational culture, which is crucial in the complex problem of DT.

Keywords: artificial intelligence, intelligent agent, expert system, knowledge engineering, knowledge engineering methodology, fuzzy logic, federated data architecture, digital culture

1.1. Part 1: Intelligent Knowledge-Based Systems, IKBS

Artificial intelligence (AI) is a way of making a computer, a computer-controlled robot, or software, think, reasoning and intelligently behave similar to humans. AI is achieved through the analogical study of the human brain's way of thinking, and how people learn, decide and work while solving a problem.

1.1.1. What AI Seeks

We are entering the 2nd century of history of the AI. The future depends on the young and brilliant minds that opt for *researching*. Previously Alan Turing had proposed his famous imitation game, which was later called the Turing Test in a paper published by the ACM (Sec. 1.1.1.9).

AI seeks to:

- Create IKBS: systems that exhibit intelligent and autonomous behavior in their decisions, or learn, demonstrate, explain reasoning and advise their users.
- Teach how to think to the machine: the creation of systems that understand, think, learn and tend to behave like people.
- As AI is a science and a technology that is based on many scientific and technical disciplines such as computing, biology, psychology, linguistics, mathematics and engineering, an important goal of AI is in the development of the computational functions associated with human intelligence, such as reasoning, learning and problem solving.
- Develop systems of high complexity with rigorous treatment: the engineering of the AI is also called *knowledge engineering* (KE) which is considered as the *expert systems* (ES) constructor.
- Add value to both an iMIS and the DT.

- Provide a system for the KS, enabling a better and greater application of modern technologies, as a step to reduce the digital divide.

1.1.1.1. Definition of Intelligence

Intelligence can be defined as the ability of a system to calculate, reason, perceive relationships and analogies, learn from experience (reason with real cases), store, solve problems, understand complex ideas, use natural language fluently, classify, generalize and adapt to new situations.

Intelligence:
• Is the systematization of knowledge • Increase the efficiency of a system in a form more than proportional • Is the highest added value for the DT and the organization • Enables the company to stand out, survive successfully and be competitive • Refers to how to do more with less: a today's challenge.

1.1.1.2. Intelligence Is an Intangible Intellectual Asset

Intelligence is composed of:

1. *Reasoning*: the set of processes that allows to provide a basis to establish judgments, make decisions autonomously and predict. In general, there are two types:
   - Inductive reasoning: its purpose is the study of the tests that allow to measure the possibility of arguments, as well as the rules to construct solid inductive arguments. Unlike deductive reasoning, in inductive reasoning there is no agreement on when to consider an argument as valid; it carries out specific observations to make broad general statements. Even if all the premises are true in a statement, the inductive reasoning allows the conclusion to be false.
   - Deductive reasoning starts with a general statement and examines the possibilities of reaching a specific logical conclusion.

2. *Learning*: is the most important property of the human being. It is the activity of incorporating, appropriating and endogenizing knowledge or skill through study, practice, education or process experience. Learning increases the awareness of the study subjects. The learning capacity is possessed by humans, some animals and systems enabled for AI.

A taxonomy of learning:

- Auditory learning: learning by listening and hearing.
- Episodic learning: learning by remembering sequences of events that one has witnessed or experienced. It is linear and orderly.
- Motor learning: learning by precise movement of the muscles. For example, choose objects, write, etc.
- Observational learning: learning by looking and imitating others.
- Perceptual learning: learning to recognize the stimuli that have been seen before.
- Relational learning: involves learning to differentiate between various stimuli on the basis of relational properties, as opposed to absolute properties.
- Visual learning: learn through visual stimuli (hyper-media) such as images, colors, maps, etc.
- Behavioral learning, stimulus-response: learning to perform a particular behavior when there is a certain stimulus.

3. *Problem solving*: process in which an objective solution is perceived and attempted, from a present situation by taking a path, which is blocked by known or unknown obstacles and restrictions. It also includes decision-making autonomously, which is the process of selecting the best suitable alternative among multiple choices to achieve an objective.
4. *Perception*: is the process of acquiring, interpreting, selecting and organizing sensory information. Perception involves detection. In people, perception is supported by sensory organs. In the AI domain, the perception mechanism gathers the data acquired by the sensors in a significant way.
5. *Linguistic intelligence*: the ability to use, understand, speak and write natural verbal and written language for interpersonal

communication. It implies the recognition and use of phonology mechanisms (speech sounds), syntax (grammar) and semantics (meaning).

1.1.1.3. Other Intelligence Classes

Researchers have identified more than 30 varieties of intelligences. To mention some of them:

1. *Intelligence and spatial reasoning*: is the perception of visual or spatial information, modify it and recreate visual images without reference to objects, build 3D images and move and rotate them.
2. *Intrapersonal intelligence*: to distinguish between one's own emotions, intentions and motivations.
3. *Relational intelligence*: as the potentiality to recognize and make distinctions between the feelings, beliefs and intentions of other people.
4. *Rational intelligence*: to understand complex and abstract ideas of *mathematics*. Rationality is a quality or state of being rational.

It is important to differentiate between human and machine intelligence:

- Humans perceive by patterns while machines perceive by set of rules (algorithms) and data.
- People store and retrieve information by patterns; the machines do it by using algorithms.
- Individuals can discover a complete object even if part of it is missing or distorted; while the machines cannot do it correctly.

1.1.1.4. Intelligent Agents

The *Ecosystem of the AI* is composed of agents and their environment with their elements. Agents act in their environment. The environment can contain other agents. An agent is any system that can perceive its environment through sensors and acts on that environment through effectors.

- The human agent has as sensory organs: eyes, ears, nose, tongue and skin; and has other organs such as hands, legs, mouth, as effectors.
- A robotic agent replaces senses by infrared and other range cameras and selectors as sensors, and several motors and actuators as effectors.
- A software agent has encoded bit streams as its programs and actions.

An *ideal rational agent* is one who is capable of performing the expected actions to maximize its performance, based on:

- Its sequence in perception.
- An integrated knowledge base (KB).

1.1.1.5. The Rationality of an Agent

The reasonableness of an agent depends on:

1. Performance metrics, which determine the degree of success.
2. Sequence of perceptions (inputs) of the agent so far.
3. The prior knowledge (KB) of the agent about the environment.
4. The actions (outputs) that the agent can carry out.

A rational agent always performs the correct action (the planned one), e.g., the action that makes the agent more successful in the given perception sequence.

1.1.1.6. The Structure of the Agent

This can be seen as:

- Agent = architecture + agent software.
- Architecture = the machinery in which an agent is executed.
- Agent software = an implementation of certain functionalities for that agent.

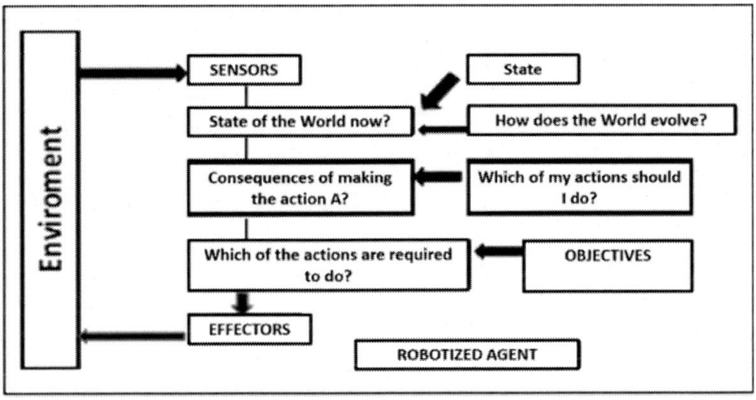

Figure 1.1. Agent structure.

Simple reflex agents:

- Choose actions based only on current perception.
- They are rational only if a correct decision is made only on the basis of a measure.
- Their environment is completely observable and identifiable.

Reflex agents *oriented by models*: use a model (i.e., functionalities or behaviors on "what to do") of the world to choose their actions. They maintain an internal state, i.e., this internal state is a representation of unobserved aspects of the current state according to the history of the perceptions.

The *status update* requires information about:

- How does the world evolve?
- What are the consequences of the actions of the agent on the environment?

Agents *based on objectives*: choose their actions to achieve the goals. The objective-based approach is more flexible than the reflex agent one because the knowledge that supports a decision is explicitly modeled, allowing modifications.

Goal: is the description of a desirable situations to reach.

1.1.1.7. The AI Ecosystem

Within *AI Ecosystems* we find some elements such as:

- DATA: It is the closest thing to reality. It is one of the multiple determinations of a problem.
- INFORMATION: It is the systematization of data. It is the input to a *decision-making process*; responds to managerial styles through organized, processed data.
- FIRM: is the *amplifier of human effort*. The company does not remain resolved, it is a "living" system. It has this as an added value to its *information system*.
- WISDOM: It is the systematization of the intelligence of knowledge. Enables transcendence as a systemic effect in the digitized organization, and also *increases* the efficiency of a digital system. It is the greatest added value for the Society facing the IV Industrial Revolution; it makes the digital company more meaningful. It is the *cultural intelligence* of the organization applying *judgments and values* to the understanding of the problems.

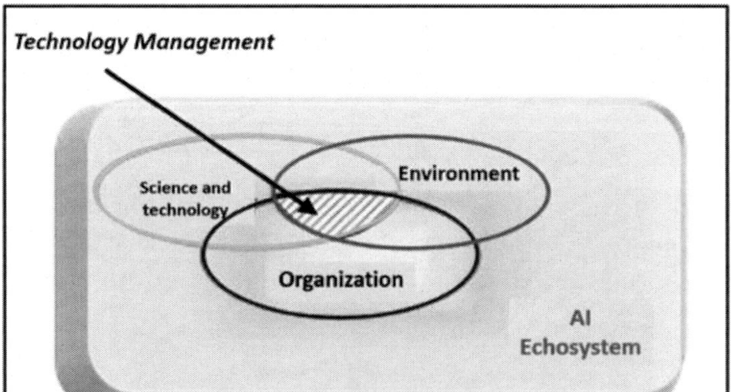

Figure 1.2. The AI ecosystems in technology management space.

1.1.1.8. Properties of the Ecosystem Environment

The environment has multiple properties:

- *Discrete vs. continuous*: if there are a limited number of distinct and clearly defined environmental states, the environment is discrete (for example, chess-plays); otherwise, it is continuous (for example, walking).

- *Observable vs. partially observable*: if it is possible to determine the complete state of the environment at each time point from the perceptions, it is observable; otherwise, it is only partially observable.
- *Static vs. dynamic*: if the environment does not change while an agent is acting, then it is static; otherwise, it is dynamic.
- *Single agent vs. multiple agents*: the environment may contain other agents that may be of the same type or different from the agent.
- *Accessible vs. inaccessible*: if the agent's sensory apparatus can access the complete state of the environment, the agent can access the environment.
- *Deterministic vs. non-deterministic*: if the next state of the environment is completely determined by the current state as well as the actions of the agent, then the environment is deterministic; otherwise, it is not-deterministic.
- *Episodic vs. Not episodic*: in an episodic environment, each episode consists of the agent perceiving and then acting. The quality of its action depends only on the episode itself. Subsequent episodes do not depend on the actions in the previous episodes. Episodic (having the form of an episode) environments are much simpler because the agent does not need to think about the future.

In the Section 4.4 we discuss more about *ecosystems*.

1.1.1.9. The Turing Test

Some programs operate in a completely artificial environment limited to keyboard input, database, computer file systems and character output on a screen. On the contrary, there are some software agents (software robots or softbots) in rich and unlimited domains. The simulator has a very detailed and complex environment. The software agent must choose between a long series of actions in real time. A softbot designed to deduce the preferences in the client's instances and show interesting elements for the client work in a real and artificial environment. The most famous artificial environment is the Turing test, in which a real agent and other artificial agents are tested under equal conditions. This is a very challenging environment since it is very difficult for a software agent to work as well as a human being.

The Turing test is a test of the ability of a machine or systems to exhibit intelligent behavior similar to that of a human being or indistinguishable from it. Alan Turing proposed that a person evaluate conversations in natural language between a human and a machine designed to generate responses similar to those of a human. The success of an intelligent behavior of a system can be measured with this Test. Two people and a machine to be evaluated participate in the test. Of the two people, one plays the role of the tester. Each of them sits in different rooms. The tester does not know who the machine is and who is a human. He interrogates the questions by keyboard and sends them to both intelligences, from which he receives transcribed answers. This test aims to deceive the tester. If the tester cannot determine the response of the machine from the human response, then the machine is said to be intelligent.

1.1.2. Expert Systems (ES)

ES are the computational applications developed to solve complex problems in a particular domain, at the level of intelligence and human experience or of an intelligent machine.

1.1.2.1. Characteristics of ES

- High performance
- Understandable and readable
- Reliability
- Highly receptive.

The ES are capable of:

- Advise
- Instruct and help in decision making
- Demonstrate theorems
- Solve structured problems
- Diagnose
- Explain a line of reasoning
- Interpret questions
- Predict results
- Justify the conclusion
- Suggest alternative options to a problem.

1.1.2.2. Components of ES

- Knowledge base (KB)
- Inference machine (IM)
- User interface (UI).

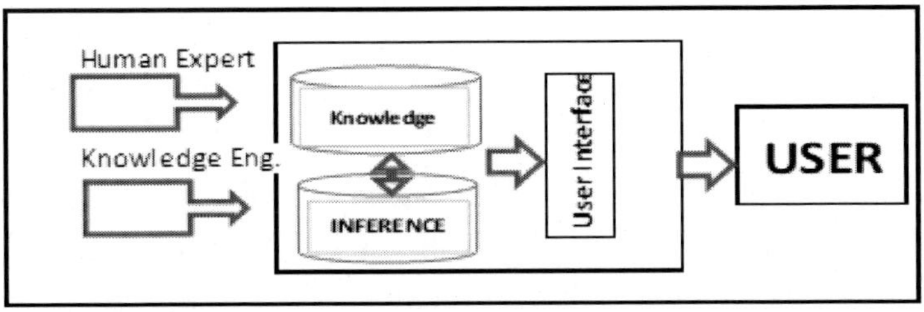

Figure 1.3. Knowledge base.

1.1.2.3. Knowledge Base

Contains knowledge of the domain. It requires knowledge processing to show intelligent behavior. The success of any ES depends mainly on the acquisition of appropriate, accurate and relevant knowledge.

The data is gathered in facts. The information is organized as data and facts about the domain of the task. The combined data, information and past experience make up the KB.

Knowledge: Instructions to do something, procedures to do a task, rules to face the complexity, problems solving. Required for business survival. It refers to the how to do? What is correct?

The incorporation of layers of knowledge:

- Offer explanations about a decision: *intelligentiation*
- Have a certain degree of autonomy in decisions: intelligence
- Offer intellectually more demanding environments as a requirement to develop superior management skills: *cognimatization*.

1.1.2.4. Components of the KB of an ES

It is the *repository* of both factual and heuristic knowledge, namely:

- Knowledge of facts: is the information widely accepted by knowledge and academic engineers in the domain of the task
- Heuristic knowledge: it is about practice, precise judgment, capacity for evaluation and divination.

Knowledge representation - *Ontology*: It is the method used to organize, interpret and formalize KB knowledge. It can be in the form of IF-THEN-ELSE rules.

The *acquisition* of knowledge: The success of any ES depends mainly on the quality, integrity and precision of the stored information in KB that is formed by the contributions of several experts, academics and *knowledge engineers*.

The KE *knowledge engineer*:

- Is a person with the qualities of empathy, rapid learning, and analysis skills of real or virtual cases.
- Acquires information from the expert on the subject when recording, interviewing and observing it at work, etc.
- Categorizes and organizes information in a meaningful way, in the form of IF-THEN-ELSE rules, to be used by the interference machine.

- It also supervises the development of the ES.

1.1.2.5. Inference Engine (IE)

The use of efficient procedures and rules by the IE is essential to deduce a correct solution. In the case of ES based on knowledge, the IE acquires and manages the KB to arrive at a particular solution. In the case of ES based on rules (as mentioned, a form of ontological knowledge representation), the IE:

- Applies rules repeatedly to the available facts, which are obtained from the application of specified rules.
- Add new knowledge to the KB if necessary.
- Resolves rule conflicts when multiple of them apply to a particular case.

To recommend a solution, IE uses the following strategies:

- Forward chaining, which is a strategy of an ES to answer the question: "What can happen next?" Here, the IE follows the chain of conditions and derivations and finally deduces the result; considers all the facts and rules, and orders them before reaching a solution.
- Backward chain or retro-reasoning, which is the strategy followed to work on the conclusion, the result or the effect. For example, the prediction of the state of the stock market as an effect of changes in interest rates.

1.1.2.6. User Interface (UI)

UI provides interaction between the user and the ES. In general, it is based on natural language processing to be used by the user who is well instructed in the domain of the task. The ES user does not necessarily need to be an AI specialist. You can explain how the ES has come up with a particular recommendation, i.e., line of reasoning, which rules are used, list of numbers of rules to be displayed on the screen. This UI makes it easy to track the credibility of deductions.

Functional and non-functional UI requirements to be effective:

- Must help users to achieve their goals in the shortest possible way.
- It must be designed to work for the existing or desired jobs of the user.
- Its technology must be adaptable to the *engineering of user requirements*.
- It must make efficient use of the user's questions.
- It must be of quality, free of errors.

1.1.2.7. Limitations of the ES

No current technology can offer an easy and comprehensive solution. Large systems are expensive, require significant development time and many computing resources. The ES have their conditions, which include:

- Technology limitations
- Difficult to acquire knowledge
- ES are difficult to maintain
- High development costs.

1.1.2.8. Variety of ES Technologies

Among other ES technologies, are included:

1. ES development environment: includes hardware and tools:
    - Workstations, desktop, mainframes
    - High level symbolic programming languages like LISP and PROLOG.
    - Large databases.
2. Tools: greatly reduce the engineering effort and the cost involved in the development of an ES:
    - Powerful editors and KB debugging tools with multiple windows

- Providers of rapid prototyping at different levels
- Variety of knowledge representation and UI designs.

3. Shell: is just an ES without the KB. A shell provides developers:
   - The acquisition of knowledge IE and the explanation facilities
   - Java Expert System Shell (JESS) that provides the fully developed Java API to create an SE.

1.1.2.9. Development Methods of ES

Generally: the ES development process is iterative. The development of the ES includes:

1. Identify the domain of the problem:
   - The problem must be adequate for an ES to solve it; e.g., a work of art is not feasible since it is not related to structured knowledge.
   - Find the KE and *experts* in the domain of the task, for an IKBS project.
   - Establish the return on investments (ROI) of the system.
2. System design:
   - Identify ES technology.
   - Know and establish the degree of integration with other systems and databases.
   - Understand how concepts can best represent knowledge of the domain.
3. Develop the prototype:
   With KB forms, the KE works for:
   - Acquire expert knowledge of the domain.
   - Represent it in the form of If-THEN-ELSE rules.
4. Test and refinement of the prototype:
   - The KE uses sampled cases to test the prototype in search of deficiencies, faults and imbalances in its performance
   - End users test ES prototypes.
5. Development and completeness of the ES requires:

- Test and ensure the interaction of the ES with all elements of its environment, including end users, databases and other information systems.
- Document the ES project well.
- Train the user to use ES.

6. Maintaining the ES implies:
    - Keeping the KB up to date by review and with periodic updates.
    - Pay attention to new interfaces with other information systems, as those systems evolve.

1.1.2.10. Benefits of ES

- Availability: ES are easily exploitable due to the actual production and complexity of the software.
- Low production cost, i.e., is reasonable that this makes them accessible.
- Speed: they offer great mobility; they reduce the work effort that a person has to make.
- Lower error rate: the error rate is low compared to the errors of people. Reduce the risk: they can work in a dangerous environment for humans.
- Stable response: they work constantly without having to move, tense or fatigue.

1.1.3. AI and KE Methodology for DT: For a Formal Methodology for the DT

Our research effort represents ongoing work with several *university projects,* unpublished graduate works, and also consulting works for institutions in Colombia. In the FESSANJOSE the model has evolved to our proposal of *social innovation using the W. Leontief model* that incorporates the knowledge sector.

The *AI and KE Methodology* presented is an attempt for reducing the complexity involved: is part of that effort and has not obeyed to a large-scale project with good funding. They have been serene developments that we want to share with the academic community. We describe an attempt to formalize a *KE Methodology* in 8 steps which are presented in detail, the mix of ideas is coming from both *engineering* and *management* that we tried to integrate in this construct. Also, we discuss briefly our understanding between *knowledge engineering* and *knowledge management.*

1.1.3.1. Background of the Method

For several years we have been involved in building ES, AI and courseware development in areas such *strategic planning* and *information systems architectures* based on knowledge and building intelligent tools for the *superior education sector*, which has resulted in an empirical study we have gradually put on several congresses.

1.1.3.2. Proposals towards an IA and KE Methodology

KE is considered as the systemization of intelligence and relies on the work of organization, representation and knowledge management as a substitute for certain tasks of human reasoning and reasonable basis to encapsulate solutions to hard problems and difficult tasks, it also serves to identify the layers of knowledge of an information system. In itself, the KE is the technology for building ES and intelligent agents. Clearly the maturity of the KE requires a disciplined work and treatment of problems, analysis and solutions, supported by knowledge.

1.1.3.3. Some Implications of Intelligent Engineering Solution

We found at extensive knowledge review in the subject or study target, that informal or no explicitness is one of the consequences; also, that the use of knowledge representation with a model of expertise and the application of search techniques in the appropriate information structure is required, as well as the support for heuristics analysis and for the same creativity.

Finally, the *main competences required* are: the ability to infer knowledge, the ability to reasoning or symbolic processing, and the ability to apply an appropriate logic of the problem. In DT the increase in business intelligence (BI) is impressive, supported by ES and intelligent systems (IS).

1.1.3.4. Engineering of the AI and Knowledge

This construct refers to the techniques of obtaining knowledge, representing, building and using it appropriately to assemble and explain reasoning of the knowledge-based system, and creating a product which tries to solve the involved problem.

The following defined steps were identified, that describe graphically the proposed methodology of KE (the two phases are related to the *problem definition* and *determination*), namely:

- *Cognitive Planning*: defining the system (structure and components), model (behavior), and detailed specification of the problem, selecting methods and techniques, user requirements and interface, identifying the data types, objects, and the solution quality.
- *Learning*: acquisition of knowledge or machine learning process; acquisition of knowledge required by the system, in a systematic or automatic way or with help from the experts. This knowledge can be specific when the problem domain and solution procedures are the same, or may be general knowledge, i.e., knowledge across an enterprise.

The subsequent 3 phases are related to the *development* of the intelligent system:

- *Ontology*, representation and organization of knowledge: how it is distributed, classified, and organized; how was developed the map of knowledge and how it is represented by some formalism or structured information or objects in the KB, to enable computer use.
- *Knowledge acquisition methods.*
- *Problem solving design* alternatives and selection of the solution.

- *Prototype construction. Verification, validation, contrasting and legitimacy*: is the methodology used successfully for the problem? Verify sources of knowledge. Legitimate and check that the knowledge is represented in the KB correctly. Demonstrate that the solution and conclusions are as expected.

The inference is based on the design of software, which allows the computer to infer on the knowledge represented in the KB. This includes:

- *Explanation*
- *Justification.*
- *Refinement*: based on the different opinions of experts, an analysis of KB and, if the IA and *knowledge engineer* finds it necessary, a redesign or restatement is realized.
- *Assessment*: there is a close relationship between evaluation and refinements of the ES as the evaluation may reveal cases handled by the rules of the system and as a result, we add new rules or modify old ones; such changes can have unexpected negative effects on some parts of the system.
- *Verifying*: the purpose of this evaluation is to check if the system implements its specifications and there is good logical consistency (i.e., the problem of verifying that the method is correct). For this purpose, the assessment is obtained, taking into account actual performance levels. The prototype and system upgrades should be tested for performance in the laboratory and in the field. The initial evaluation is conducted in a simulated environment, in which the system is exposed to test problems (e.g., case histories or suggested by users). It also examines whether the system can be used efficiently and if it is cost-effective. The assessment should consider the quality of the messages presented, i.e., design and programming of the explanatory power of the KB to determine to what extent it meet the proposed objectives.
- *Validation*: should test the system and application performance, which makes comparison with experts, to establish that the system

built to solve the problem is correct and will perform at an acceptable level of accuracy. The decrease in response time and quality of the conclusions may be a good initial approach to evaluate an ES. For this purpose, we use the *modified Turing test*, with which two solutions to one problem are presented to potential users or administrators: the first is the result of expert opinion and the other the result of an iMIS. Without knowing the source, the persons are asked to compare the solutions. The results of the comparison determine how valid the results of the iMIS are. To use this approach, it should be considered the potential disagreement between the criteria of the evaluators, since the problems can be so complex that occur disagreements over their interpretation and solution.

Other requirements to be considered in *validating the final system* are:

1. Should meet a recognized and important business assistance, or counseling or tutoring need.
2. The processing speed of the system must be very high.
3. The system must be able to increase the skills of the user: dual learning, man-machine.
4. Error correction should be easy to perform.
5. The system must be able to answer simple questions.
6. The system should ask questions for additional information.
7. Knowledge should be easily modified (allow add, delete and modify rules.)
8. Make the user feel control the system to encourage its use.
9. It must be a reasonable degree of physical and mental effort for a novice user.
10. Must be clear requirements in terms of data entry and this should be simple to obtain.
    - Obtaining the *final system and evolution*: one advantage of rule-based systems is that they are modular, so can build subdivisions of large systems and then test them step by step,

which is made possible with prototypes. One approach is using the evolutionary prototyping, where a prototype is enhanced until obtaining the final system, which can be built gradually by adding pieces, parts or modular.

- *Component*: this is a cyclical process, which is an advantage that can improve the result and get the best approach to user requirements and needs of iMIS knowledge. If each subsection is evaluated and approved separately, the final system will most likely work the first time. In this sub-step integration of subsystems is performed, that is, if KB is created different or heterogeneous in terms of sources, we must take into account the need to create interfaces between them, allowing optimal operation and communication between the system and users.

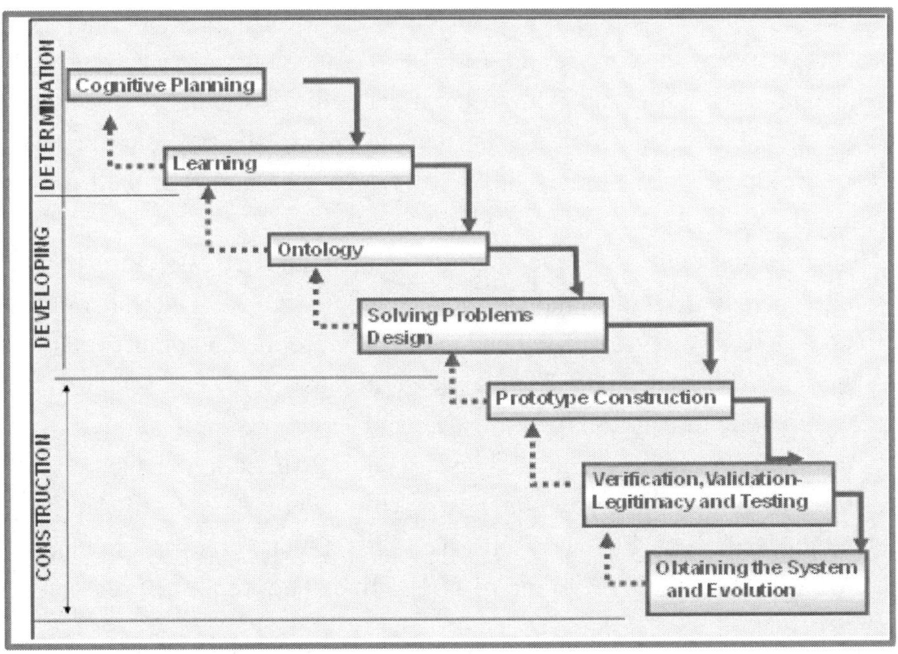

Figure 1.4. AI and KE life cycle.

1.2. Part 2: Basic and Conceptual Aspects for Digital Transformation

1.2.1. Fuzzy Logic Systems

1.2.1.1. Introduction

A conceptual framework is presented on the *fuzzy logic systems* (FLS), which we have used in our constructs. Our proposal developed extensively the *federated data (FD) Matrix*, that is presented with a synthesis of its use and application in HE, and also in professional contexts.

1.2.1.2. Fuzzy Logic Systems

These are systems that produce admissible but defined results in response to partial, ambiguous, incomplete, distorted, undefined or inaccurate (fuzzy) data. *Fuzzy logic* (FL) is a method of processing that resembles human reasoning. The FL approach mimics the path of decision making performed by individuals that involves all the intermediate possibilities between the binary or digital values YES and NO; the truth is fractioned.

The conventional logic block (traditional mathematical logic) that a computer can understand takes a precise input and produces an output defined as TRUE or FALSE, which is equivalent to the human affirmation or negation, YES or NO. The *Boolean algebra* has its roots for its mathematical treatment in large industrial and economic applications.

Ambiguity, or logic of incomplete reasoning, coined in our medium as FL, is based on the *fuzzy sets theory*, creation of professor Lotfi Aliasker Zadeh, the father of FL, who observed that unlike computers, human decision making includes a range of possibilities between YES and NOT, such as:

YES
POSSIBLY YES
I CANNOT TELL

POSSIBLY NOT
DO NOT

FL works from the levels of several input possibilities to achieve the defined output.

1.2.1.3. FLS Implementation

FLS can be implemented in:

- Systems with various sizes and capacities, ranging from small microcontrollers to large network control systems based on workstations.
- Hardware, software or a combination of both.

1.2.1.4. Purposes

FL is useful for commercial and practical purposes; it:

- Can control machines and consumer products.
- May not provide accurate reasoning, but acceptable reasoning.
- Helps deal with uncertainty in engineering.

1.2.1.5. Fuzzy Logic Systems Architecture

FLS have four main parts as shown:

1. Fuzzification module: transforms system inputs, which are clear numbers, into fuzzy sets. Divide the input signal in five steps, such as:
   GV x is great true
   MV x is true medium
   P x is small
   MN x is negative medium
   GN x is big negative
2. Knowledge base: stores IF-THEN rules provided by experts.

3. Inference Engine: simulates the human reasoning process by making diffuse inference about the inputs and IF-THEN rules.
4. FLS Module: produce acceptable but defined results in response to incomplete, ambiguous, distorted or inaccurate (fuzzy) data.
   $\mu A: X \rightarrow [0,1]$.

   Here, each element of X is assigned with a value between 0 and 1. It is called membership value or membership grade. Quantifies the degree of belonging of the element in X to the diffuse set A:

   - x-axis (rows of the fuzzy dashboard matrix, FDM) represents the universe of discourse (productions)
   - y-axis (columns) represents the membership degrees in the interval [0, 1].

There may be several applicable membership functions to blur a numerical value. Simple membership functions are used because the use of complex functions does not add more precision to the output.

Triangular membership function forms are more common among other forms of membership function such as trapezoidal (Figure 1.5), singleton and gaussian; there, i.e., the input to the 5-level fuzzifier varies from -10 volts to +10 volts. Therefore, the corresponding result also changes.

1.2.1.6. Algorithm

1. Define linguistic variables and terms (start)
2. Build membership functions for them (start)
3. Build knowledge base rules (start)
4. Convert sharp data into fuzzy data sets using the membership functions (fuzzification)
5. Evaluate rules in the rule base (inference engine)
6. Combine the results of each rule (inference engine)
7. Convert output data to non-fuzzy values (defuzzification).

1.2.1.7. Advantages of the FLS

- Mathematical concepts within fuzzy reasoning are simple as seen with the FDM. FLS are easy to build and understand.
- You can modify an FLS simply by adding or deleting rules due to the flexibility of fuzzy logic: e.g., the number of *experts*, the granularity of the fuzzy values, the colors, etc.
- Fuzzy logic is a solution to complex problems in many areas, since it resembles professional reasoning and decision making.

1.2.1.8. Disadvantages of FLS

- There is no invariable approach to the design of fuzzy systems.
- They are understandable only when they are simple.
- They are suitable for problems that do not require high precision.

1.2.1.9. Complements

A strategy to present the fuzzy predicate logic strategy consists of remembering some aspects of the theory of conventional sets (which we will call concrete sets), and from there to make an extension to the fuzzy sets:

- A specific set is defined as a repertoire of elements that exist within a *reference universe*. It is based on estimating the membership or belonging to a set (or system).
- The membership function associated with concrete sets can only have two binary values: 1 or 0 (YES or NOT, it is a member or not, things are or are not, belong or do not belong); while in fuzzy sets (fuzzy logic) can have any value between 0 and 1, i.e., they may belong partially; in Figure 1.5 they are shown with the striped area of the trapezoids that represent sets. In fact, reality is diffuse.
- An element can belong (partially) to a diffuse set and simultaneously belong (also partially) to the complement of the said set (see shaded area, fuzzy zone in Figure 1.5). The previous thing is not possible in the concrete sets, since it would constitute a violation at the beginning of the excluded third.

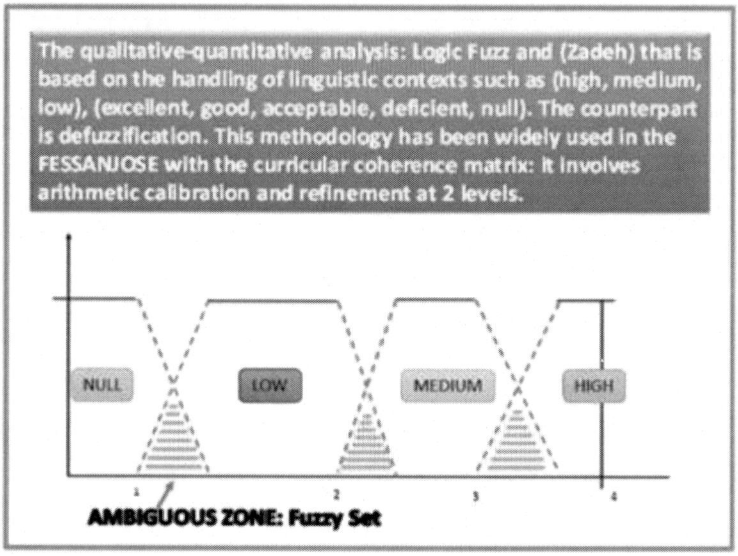

Figure 1.5. Fuzzy space.

- The boundaries of a specific set are exact, while those of a diffuse set are diffuse, since there are elements in the borders themselves, and these elements are both inside and outside the set. This is the case of linguistic contexts: good to bad ranges.

Table 1.1. Correspondence between operators of the theory of sets, the Boolean algebra and the traditional logic

Theory of Sets	Traditional Boolean Algebra	Logic
Intersection	Conjunct	AND
Disjunction	Union	OR
NOT	Complement	DENIAL

It should be noted, on the other hand, that it is well known that set theory, Boolean algebra and mathematical or traditional logic are isomorphic, under appropriate transformations. Isomorphism refers to a similar underlying structure, and therefore the definitions that are made in any of the three theories can be taken to the other two, through appropriate transformations. Table 1.1 shows the correspondence of some operators.

1.2.1.10. Precedents of our Fuzzy Dashboard Matrix (FDM)

As mentioned above, the fuzzy predicate logic was proposed by the pioneer, Prof. Lofty Zadeh, from which our constructs emerged in the FESSANJOSE Research Group:

1. For the analysis of curricular coherence and its quantification: the curricular architecture integrated to the methodology of the *propaedeutic cycles* (*professional technician, technology* and *professional*) of the Ministry of National Education (MEN[8]) for a *plan of studies*. It is based on 3 dialogical components: the structural or *Matrix Model* for curricular coherence analytics[9], the *Knowledge Management* and the *Student Productivity Model.* It makes the curricular analysis possible in the time domain (periods) and in the domain of knowledge (by areas and subjects); and measures the student cognitive production at the *university*.
2. Following the Convocation 745 of COLCIENCIAS[10], resuming research and previous experiences of the 755 MEN - FESSANJOSE Agreement (2011-2013): with the idea of evidencing and quantifying the relevance of our proposals with the architecture of our compendium or portfolio of final and intermediate products, against the objectives, sub-objectives, lines and sub-lines, using *fuzzy* logic of AI with qualitative values. The construct makes possible to do quantitative and qualitative analysis of portfolios. These developments constitute the qualitative and quantitative starting points to explore extensions and potentialities of the objectives, quantitative goals, sub-objectives, impacts, incidents, etc., of the portfolio, tending to develop projects focused on the knowledge society in the 3rd Millennium, inherent to the *high*

---

[8] www.mineducacion.gov.co/

[9] Book Chapter: Analytical Models for Tertiary Education by Propaedeutic Cycles Applying Knowledge Engineering and Knowledge Management. Title: *New Research on Knowledge Management Models and Methods*; published by INTECH - Open Access Publisher covering the fields of Science, Technology and Medicine, ISBN 979-953-307-226-4.

[10] *Colombian Administrative Department of Science, Technology and Innovation*, www.colciencias.gov.co.

quality of the portfolio. The construct is implemented in FESSANJOSE-Bogotá with all the documentation under the IEEE standards, also including a *handbook* for *system operation*: USANJOSE-PORTFOL-ANALYSIS.

1.2.1.11. Design Specifications: FDM Matrix Architecture

In general, this matrix A is MxN where:

- The M rows represent the levels of the purposes and objectives to be obtained from the *portfolio*.
- The N columns represent the unitary and functional areas of *the products, goals*.
- Each coefficient (cell) $a_{i,j} = \{0, 1\}$, for the case of a binary matrix, is 1 if there is direct involvement between the area of expertise and each subject of the curriculum of the cycle, and is 0 otherwise.

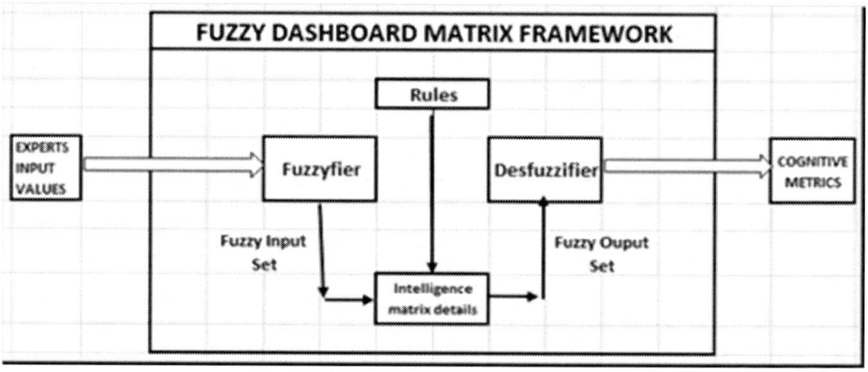

Figure 1.6. Fuzzy FDM dashboard matrix.

On the other hand, each coefficient (cell), which represents the cohesion and coherence of each objective with production, can contain other qualitative wider values (fuzzy logic), so that the degree of possibilities is increased:

$a_{i,j}$ = *{null, low, medium, high} (display representation: white, green, red, blue in the* Table 1.2*).*

Table 1.2. Fuzzy qualitative values

fuzzy values	Scalar
NULL	0
REGULAR	1
GOOD	2
HIGH	3

The rows represent the outputs quantifying the total production i°, and the columns the inputs quantifying the added value of the variable associated with column j°.

As shown in Table 1.3, there are many uses and applications of our FDM for the broad academic field, the academic digitalization in specific, the quality analysis, and the evaluation required for the DT.

1.2.1.12. Some Applications and Uses of the FDM

In the Table 1.3, PSEd Productivity, we synthesize most of the tools we have developed regarding DT in our research group.

Table 1.3. Annex: Our developed tools

A1. USANJOSE-CURR-COHE: system to analyze and quantify the curricular coherence of the plans of study of the propaedeutic cycles in higher education. Certification of logical support registry - software Book 13-Volume 41-Part 205, National Direction of Copyright, Bogotá. *Description*: curricular architecture integrated to the coherent and cohesive study plan of the propaedeutic cycles: professional technician, technology and professional. It is supported in 3 dialogic components: Structural or Matrix Model of Curricular Coherence, Knowledge Management and Student Productivity. Enables curricular analysis in the domain of time (periods) and in the domain of knowledge (by areas and subjects). Measures the student cognitive production of the university.

Table 1.3. (Continued)

A2. USANJOSE-ADVERTOR-HE: Early Warnings System. Registered software certificate logical support registration: Book 13 - Volume 41 - Part 203, National Directorate of Copyright, Bogotá. *Description:* early alerts characterizes risk according to SNIES[11] and detailed monitoring of students who have a high risk of desertion; it also sees teaching and students' compliance, and takes a daily logbook by electronic means, that gives an indicator of absence for a period of time, which prevents deterioration in academic quality; finally, it makes an individual monitoring and accompaniment about the student academic performance and the payments. The system has models of student and the teacher. **A3. SANJOSE-HE-MIS**: information system for architectures of higher education, academic and administrative management. Certified registry of software: Book 13- Volume 40- Part 307, National Directorate of Copyright, Bogotá. *Description:* complex architectural system for solution of academic - administrative problems in higher education; accounts for 5 subsequent architectures or general modules that are: academic module, collection module, student portal, teaching portal and administrative portal. Contains a confidential and scalable software business architecture developed in Visual Basic and Visual Basic .NET, with framework 4.0 motor SQL - server. Its purpose is academic– administrative management. **A6. USANJOSE-ARQ-COHE-CURRICULAR**: Intelligent Architecture of Curricular Knowledge and Competencies for the optimization of a Study Plan. Registered certified software, National Directorate of Copyright: Book 13 - Volume 45- Part 200, Bogotá. *Description:* curricular architecture integrated to the coherent and cohesive study plan of the propaedeutic cycles: professional technician, technology and professional. It is supported in 3 dialogic components: Structural or Matrix Model of Curricular Coherence, Knowledge Management and Student Productivity. Enables curricular analysis in the domain of time (periods) and in the domain of knowledge (by areas and subjects). Measures the student cognitive production of the university. **A7. USANJOSE-CUR-COHERENCE-ARCHITECTURE**: Intelligent Architecture of Curricular Knowledge and Competencies for optimizing a Curriculum in Higher Education. Registered software certificate, National Directorate of Copyright: Book 13 - Volume 45 - Item 197, Bogotá. *Description:* curricular coherence architecture is linear mathematical and analytical tool of academic information, enabling the curriculum making sense as a whole complex. Academic coherence refers to logical-cognitive relations between the parts of the architecture of an educational system. The tool is an instrument of analysis which provides the optimization of an engineering curriculum using propaedeutic cycles, and weaves the coherence and consistency both internally and externally.

[11] *Sistema Nacional de Información de la Educación Superior* [*National System of HE Information*], MEN, Colombia. https://www.mineducacion.gov.co/sistemasinfo/SNIES/.

A6. USANJOSE-ARCH-MICRO-MACRO-SKILLS: intelligent architecture of curricular knowledge with unbundling of micro-competences and competences objective for the optimization of a curriculum. National Directorate of Copyright certified registered software Book 13-Volume 58- Item 16. Bogotá, Jan. 17, 2017.
Description: the construct emerged as a result of the Call 745, with the idea of extending the applications of the Curricular Coherence Matrix, in the macro competence space - (objective) versus the knowledge of the curriculum, in which they are specified for each subject, the micro-competences to cultivate in the student, it is possible to analyze and quantify the added value of the micro-competences to the methodological set of the Study Plan in its condition of exogenous variables. The developments constitute the qualitative and quantitative starting point to explore extensions and potentialities of competences in terms of skills, abilities, expertise, proficiencies, capabilities, etc. tending to develop curricula focused on the knowledge society in the 3rd Millennium, inherent in the High Quality of Higher Education.
Implemented by FESSANJOSE in Bogotá. Handbook - title: Compendium of Intelligent Architecture of Curricular Knowledge with Disaggregation of Micro-Competences and Competencies-Objective Including Academic Guidelines for Optimization of Study Plans; ISBN 978-958-48-0410-5.

A10. USANJOSE-PORTFOL-ANALYSIS: complex architecture supported by the fuzzy logic for quantitative and qualitative analysis of the product portfolio and its evaluation. National Directorate of Copyrights, certified software Book 13-Volume 58- Item 17. Jan 17, 2017.
Description: the construct emerged as a result of the Call 745, with the idea of demonstrating and quantifying the relevance of our proposal with an architecture of our compendium or portfolio of final and intermediate products in front of the objectives sub-objectives, lines and sub-lines of the call, using the fuzzy logic of artificial intelligence with qualitative values. It allows to make quantitative and qualitative analysis of portfolios. The developments constitute the qualitative and quantitative starting point to explore: extensions and potentialities of competencies in terms of skills, abilities, expertise, proficiencies, capabilities, etc.; tending to develop curricula focused on the knowledge society in the 3rd Millennium, inherent to the high quality of higher education.
Implemented by FESSANJOSE in Bogotá. Handbook - title: Compendium of Complex Architecture Supported by the Fuzzy Logic for Quantitative and Qualitative Analysis of Portfolios of Products and Results with Management Guides for Evaluation of Relevance; ISBN 978-958-48-0411-2.

A11. USANJOSE-SELF-EVAL-ANALYTICS: intelligent model of institutional self-assessment. Alfonso Pérez Gama registered it in the DNDA in Book 13 - Volume 65 - Item -52. Dec. 26, 2017.
Description: it's a complex architecture supported by the fuzzy logic of the artificial intelligence for quantitative and qualitative self-assessment analysis of research and innovation products, facing different dimensions and multiple determinations of knowledge management research.

A12 USANJOSE-SOCIALPROJ-ANALYTICS: Intelligent model for University Social Projection and external and international relations. The social projection of the university is one of the fundamental pillars of our DT model. It requires the mapping of products, services, knowledge, consultancies and so on, on society, industry, government and even the rest of the world. In synthesis, it refers to external relations, which obviously includes internationalization.

1.3. PART 3: ORGANIZATIONAL CULTURE

1.3.1. Culture

The organizational culture (OC) gives its identity to an institution. It includes the values, beliefs and behaviors of life itself. It is considered as the way in which activities are carried out in the organization. Therefore, it constitutes a bulwark if it is aligned with DT strategies. OC is formed according to the decision-making style, and especially, in the governability model, is based on the corporate principles that have been incorporated. The OC is learned and internalized, and is known through the behaviors and performances of the organization members. In special ceremonies the incentives are given to further root the OC.

The communications, whether formal or not, are pedagogical communications: circulation of values and beliefs. The level of internalized OC must be a means, regulated through surveys, monitoring of conflicts reduction, security problems, making direct observations, etc.

1.3.1.1. Developing the OC

The *development* of a culture requires its cultivation: it needs resources, it takes time, energy, and a lot of institutional and personal commitment. In order to increase this culture, a strong outreach program must be carried out through training (internal and external), workshops, circulars, email, social networks, and marketing the measures with short and effective messages.

1.3.1.2. Users

In the organization, the users are of internal and external nature:

- Clients are not always right, but we need them, and it is not necessary to win an argument.
- Users are not aware of the costs of systems.
- There is lack of commitment of the officials.

- Users have difficulties of acceptance, adaptation and use, due to lack of training and dissemination of digital meta-culture and new extreme technologies such as IOT (Internet of everything), *big data*, AI, etc.
- They are not aware of the responsibility that their management means, e. g., of the passwords; exposing the institution to social engineering and its possible implications in computer frauds.
- Lack of metrics and self-control policies.
- Lack of technological culture 4.0.
- Technophobia (fear of technology and/or its failure) aversion to use, disregard for these tools for solving problems and improving their productivity.

1.3.1.3. Cognimatics Culture

The old informatics (in Spanish: *informática*) word was a neologism derived from the words *information* and *automation* and already constituted a scientific discipline that included its own methods of validation and legitimation. Computer science in the 3rd. Millennium is already part of the basic culture of any individual in today's society. In a general sense, *computer science* was understood as the set of techniques and knowledge aimed at the transformation, recreation, and logical, computational and automatic management of data and information for better decision making and problem solving.

Throughout history, the world has suffered various technological revolutions related to information and knowledge. Associated with the Knowledge Society we have been developing *cognimatics (cognizance-automatic)*, which is the equivalent in computing to digitally process *knowledge*. These revolutions and eras have impacted in such a way that they have transformed and reorganized all practices in the economy and society. Due to the change of era: towards the IV th Industrial Revolution, the need to promote OC is generated imperatively, therefore the technological changes generate the creation of a new cognitive culture.

The culture in general, has always been in the work and in human thought, with which each individual act with the social and technological environment in diversity of situations, co-determining in the present, and in part to the future. On the other hand, the Cognimatics Culture, is what the individuals do to transform the data, information and knowledge that have in front of them, and turn these into future construction, intelligent systems and robotic constructs, sharing values and being based on educational actions. Thus, the transit to Society 4.0 becomes more viable, less traumatic, and the pillar and sustainability of the DT.

1.3.1.4. Cyber-Culture Security and Cognimatic Control

Security and how to survive are basic instincts of people, and organizations do not escape to this principle. Security problems are almost always associated with human error (due to ignorance or omission), but in a more than proportional way by the same technology. On the other hand, the enemy can be inside the organization. In most cases of computer frauds, these are carried out with internal support (*compadres* resources). In many cases, organizations develop expensive systems mechanisms to isolate themselves and protect themselves from external attacks (based on technology or not) and forget the internal threat.

According to International Data Corporation (IDC, www.idc.com) in 2017, cyber-attacks in Latin America have grown between 30% and 40%. (20 violations x second). Colombia was affected by 5 million cyber-attacks in 2017, according to the MinTIC, and the CONPES document 3854 of 2017 establishes new actions for state entities (within the e-government) and requires information shielding with the purpose of providing a better service guided by good international practices, *pari passu* with good governance of the technological infrastructure. It is very simplistic to believe that the problem is reduced to having an antivirus! The encrypted systems also do not solve *all* the problem itself. In any case, any effective countermeasure contributes to the formation of a cyber-culture 4.0.

Cognimatics is the heart of the organization and cybersecurity is aimed at guaranteeing the fundamental principles such as:

- Confidentiality and property: protects information and all intellectual assets from an interception or unauthorized knowledge, which produces the sense of ownership and belonging to the organization.

- Availability and usefulness: it allows information and all intellectual assets in addition to vital services, to be available when they are needed and useful.
- Integrity and authenticity: guarantees the accuracy and complete reception of information and all intellectual assets. Prevents the modification of information, knowledge or software in an unauthorized manner.
- Access and control: guarantees accessibility to information, to all intellectual assets and to a system. It must have the ability to limit who or what system can access, from where, at what times, what resources to use, what data to transfer, i.e., have total control of the system.
- Principle of non-rejection and audit: establishes the certainty that an operation has been performed by a valid operator. The system must maintain evidence of the use, behavior and content of each user's activities.

In short, the objective of cognitive security is to properly handle cyber-attacks and protect digital resources from damage, kidnapping, alteration, theft and loss. It is very serious that the sensitive information of an institution reaches the wrong hands given that the fact is not only the economic losses but also the systemic consequences. This cognitive security covers a wide range of strategies and solutions, among which are the following (although it is not an exhaustive list):

- *Access Control:* entry control to the system or architecture.
- *Discretional Access Control*: control of access to resources such as files, data architectures and programs, once it has been admitted into the system.
- *Planning and Administration of the System:* Planning, organization and administration of digital services, as well as policies and procedures to guarantee the security of resources.

- *Security Based on Knowledge, Logic and Physics:* to ensure the continuity of the services and the operation of the equipment.
- *Biometrics:* Use of unique characteristics to identify users.
- *Viruses and Related:* Different classes of viruses and other destructive programs, and how to prevent and control their effects.
- *Encryption:* Information cryptography must be a procedure used to preserve the information of unscrupulous users, thus allowing its use only to authorized users.
- *Security in Networks and Communications:* Security problems in communications through networks and communication systems.
- *Blockchain:* Technology that enables the transfer of digitized data with highly sophisticated coding in a completely secure way.

1.3.1.5. From the Myth of the Power of Information to the Power of Knowledge

The development of informatics in Colombia was characterized by the support of elite computer multinationals that eventually extended the computerization of our society. In the present millennium the Colombian Government (GOC) created the ICT Ministry, MinTIC. We have moved between techno-phylia and technophobia; the first enunciates the end of the kingdom of necessity, in which technological change will bring prosperity and better living conditions. The second, little trusts technology, e. g. the computer is good as a tool, data repository, complex calculator and the best typewriter: but nothing more. But the fear of cyber-war is frightening in certain circles.

The truth is that much has been written about the power of knowledge, however the reality of the socio-technical environment seems to reveal the opposite. Just the transfer of reason and thinking to the computer skills does not have the disadvantages of human distraction.

The problem is not simply not knowing how to do things: it is the opportunity to process all the available information and decide quickly why the machine is ahead of man in the face of panic, lack of control, and erratic improvisation.

1.3.2. From the Information Society to the Knowledge Society (KS)

A reflection is made in the transition from the information society to the knowledge society, characterized by the focus on scientific and technological knowledge and from this to the intelligence society of the Third Millennium, starting from the critical analysis of our own reality and of the realism of computing possibilities in the social and technical environment. Much of what we affirm here is based on our own experience. By way of synthesis, we argue that social computing is a pale reflection of what is done in Colombia at a professional level; we also suggest the production of social software for public use, as a social good.

1.3.2.1. Introduction to KS

Traditionally, the informatics appeared in the sphere of economic production in the more advanced tertiary sectors such as banking and commerce and was virtually unknown in the sphere of social production. Paradoxically it has been recognized in the computer as the most important technological change of the last century and much has been written about its intrinsic power, predicting a priori very important changes in living conditions and in social and economic progress. Today in transit to industry 4.0, more dramatic changes are coming, especially the improvements in productivity with the cyber-physical constructs.

For example, reflect on the social consequences of the tractor would give to write a few pages; it would not have much to talk about the science of the tractor either. However, the computer sciences have drawn their own knowledge and today an important space in all disciplinary knowledge, so it is essential to study the phenomenon of the computer not only as a mass media of social communication but as something indispensable in daily life, which has given rise to institutions such as UNESCO to hold important global meetings on the subject. It should be noted the Summit of the Information Society in Geneva in 2002. Generated unemployment? And, Robotization technology job killer? What dangers can represent in early childhood education? What are the social and cultural implications in individuals, in society and in government? And above all, how is the country and who is going to be the architect of the socio-technical change?

These consequences in sectors such as Culture, Education, Health, Justice, Work, Recreation and also the same Research process, are in which their own contents and methods are redefined. The above is intuitively obvious to the extent that two major trends are known in terms of computer applications: horizontal and vertical, where the first fertilizes the different disciplinary compartments, where the ubiquity of the systems makes the computers essential; vertical development refers to the vertiginous advance where third and fourth world countries do not fully know a technology when it has already completed its obsolescence cycle planned by the countries generating knowledge. This generates tremendous uncertainties in society and especially in the management of companies.

1.3.2.2. Fundamental Premises

The issue of intelligence has worried man from his very existence, today, after recording 5 generations of information technology, several things have been revealed:

- That man lost his exclusivity as a thinking being, with the new technologies of artificial intelligence and knowledge engineering, whose maturity has been highlighted and proven in advanced countries, where man artificially incorporates into the computer cognitive abilities such as reason, think, see, learn, solve problems and make decisions with a certain degree of autonomy. It causes hilarity to border the boundaries between the living, the natural and the artificial.
- The new role of the computer with machine learning, allows us to abandon our traditional trades as calculator, data repository and typewriter to assume new functions as a cognitive tool, mental mirror and intellectual colleague. Something that impresses is the real possibility of building systems with learning and dual man and machine actions which translates into increases in productivity in a significant way.

- There is a new space for the production of intelligent systems (in times of peace and in times of war), with autonomy in their decisions: mente-facts, brain-facts, techno-facts, cogni-facts, smart bombs, autonomous vehicles, etc.

1.3.2.3. Education and Change

The foregoing reveals the importance of intelligent computing and the construction of *educational mente-facts* for the socialization of knowledge and cognition of *learning* not only for theoretical-practical imperatives but for ethical and moral reasons, given that the dignity of man lies in his intelligence and creative of freedom, integrated into the other human conditions, especially affectivity and moral values.

The modifiability of cognition is not questioned, however, doing so, considering only genetic and hereditary factors would not have much room in the 3rd Millennium; doing so, supported by educational and cognitive actions has great possibilities. In this sense, several decades ago the scientist R. Thinkomorov had already pointed to the computer as the amplifier of human intelligence.

Education has taught us that changes are going to continue to happen, but the serious thing is how to learn to adapt quickly. The GOC through MinTIC promotes *change*, but paradoxically it keeps the brake by reducing budgets for the digitalization of institutions. Many users do not want change because they are already accommodated in their comfort zone, but they also deny that they need to be more productive and competitive. They are not aware that the digital age, charges for resistance to change in terms of not taking advantage of new opportunities. In addition, the new era of the IV Industrial Revolution comes with a major deceleration of more traditional institutions.

Education should train people to develop superior mental abilities with the support of artificial-computational intelligence and robotics: hybridization *human intelligence* with *artificial* and *computational intelligence* with *robotics*.

1.3.2.4. User: Front of the New Computer Role as a Cognitive Tool

In the client the expectations are greater, and it is built thinking always in the next time.

1.3.2.4.1. Experiences

What is being done to create in the customer a good experience so that the next time, the consumer wants to choose us as their supplier and not the competition? The same client is helping to develop a culture of user service and a conception of loyalty within the digital ecosystem.

It is imperative to understand that the user's knowledge and profile is indispensable today: to know their expectations, real and potential needs, their successful experiences or not. The extreme technology (e.g., IA, IOT, *big data*) and social media make the construction of user profiles more viable to make an individualized marketing with *ex ante* knowledge of it. We must understand that digital users expect service interactions in real time.

Cognimatics allows a new space for socialization. The customer experience becomes an element that enables the optimization of the service and management; this is required:

- Identify the expectations of the digital customer.
- Design and implement experiences to fulfill the promise of our brand, duly planned.
- Communicate the promise, train employees with the appropriate technology, and it means that it also requires clarity in the promise.
- Disclose in the environment the experience designed for its correct implementation.
- Design experiences that are distinctive, with multichannel in such a way that they differentiate us from the competition.

1.3.2.4.2. Other Responsibilities of the User

- Need to be updated and right to be permanently and significantly re-trained.

- The user must be aware of digital and cognitive costs.
- Directors should also be aware of the possibilities of significantly improving productivity with well-trained and digitally cultured employees.
- In the same sense as before with the clients involved, learning from their experience, identifying their expectations.
- Need for your commitment from the employee to fulfill the client as promised with our brand.

1.3.2.4.3. Values

In the process of the *strategic planning*, defining the *corporate principles* is fundamental, because they are the frame of reference of the same. There are values in different fields, such as the economic, technical, social, political religious, etc., which besides existing in each of the men, also exist at the business level and become part of their culture. The fact that values and beliefs are part of the said culture is achieved through an effective leadership that contaminates the rest of the organization, which affects their habits of thought and the way they carry out their work.

1.3.2.4.4. Importance of Values for the Strategy

The reasons are:

- The values must be appropriate for the time, place and conditions in which an *organization* works.
- The values must be immersed in the culture of the institution, OC, through the leadership that will be responsible for transmitting them to the community in the digital ecosystem.

1.3.2.4.5. Definition of Values

They are the expression of corporate philosophy, being the highest point where the purposes and goals are generated to achieve the objectives. The values are focused on purposes, and these on quantifiable goals, and goals on objectives. The corporate culture and its beliefs would basically be the method by means of which the values, purposes, goals and objectives are transmitted, and through the strategies to satisfy all this scale.

Examples of identified values are:

- Transparency supported by the quality of information and software
- Long-term goals that challenge system personnel
- Tenacity and perseverance in the face of the difficulties of the *office*
- Courage in the face of serious contingencies
- Promotion of self-criticism and self-control
- Leadership
- Respect
- Enthusiasm and self-motivation
- Commitment
- Excellence
- Honesty.

In short: the organizational culture or better the digital culture is a critical factor for the success of the DT.

Conclusion

We have presented the indispensable elements to understand the DT in terms of IKBS, ES, the methodology for KE and the OC.

We have presented basic conceptualizations of AI, intelligence, *agents* and IS with our KE methodology, which we use in our constructs. In the same way we make a discussion of *fuzzy systems* and their mathematics, with a special application: the FDM that we have used for the construction of our tools, of which we include an *Annex* of the main tools developed and registered in different instances.

The Digital Culture is one of the very important components of DT for which we dedicate this space, to establish its scope and practices. We have introduced our term: ***cognimatics***, that follows *informatics*.

Chapter 2

THE BATTLE FOR THE FUTURE

ABSTRACT

The MEN uses to design and implement policies, plans, programs and projects related to HE in Colombia (postsecondary, PSEd), in order to facilitate the access of young people to this educational *superior* level, what allows that the country has productive, trained, and committed citizens, not only with their development of their competencies, but also with the socio-scientific-technical progress of the country within the framework of a democratic society.

Building the future requires relevance for innovation and competitiveness; to promote, in the HEI, conditions to strengthen the national capacity to generate knowledge through *pari passu* research, strengthening the investigative capacity of these institutions, making the HE empowered of projects and solutions for the regions and the country.

A very important strategy is the strengthening of the offer and quality of master's and doctoral programs, in which new knowledge is produced. This requires international cooperation and techno-scientific exchange that leads to the internationalization of HE, and the mobility of students, teachers and researchers that increases highly qualified talent.

The above strategies complement each other raising the level of training of HE teachers, for science, technology and innovation in the sector and in the country.

New models of university-business-government relationship are required.

The DT is key in this construction of the future.

Keywords: MIR matrix, end-means matrix, DT objectives

2.1. Tools, Models and Architectures

PSEd means the *integration* of all higher education levels, the best tool for income redistribution, and the base for the *economics* and *technology* progress. Several research projects have been preceded with numerous international papers published (indexed in *databases* as Scopus), as well as various intellectual assets and software tools (certificated by the National Directorate of Copyrights, DNDA).

Our book tries to fill gaps that occur when the uses, applications, and constructions with the latest technologies of the III Millennium, are very scarce, and that on the other hand the problems that affect the social sectors like Education, are qualitatively and quantitatively very large, with *high* complexity, high costs and high demand for large sectors of the society and industry.

What are the possible origins? In summary, we can establish:

- Bad *competence* management. Low *academic productivity* of teachers and students; *little* importance of *technical & technology of education*; *poor* investment in RD + I.
- Poor desertions early *alerts and road maps* in student permanence and in also, reduced *education quality policies*.
- The poor solutions that have been implemented with unsatisfactory results. The recent encouragement to postgraduate teachers is insufficient because it favors only a small part of the population.
- *Bad education governability*. Deficiencies in information and indicators, infrastructure and *information system*.

2.2. Mission Critical Problems in PSED

The identified problems are:

- No articulation of the *middle education* (ME) with the HE.
- No *high quality assurance* at any level which is aggravated due the use of traditional *educational models.*
- High student desertion at all levels. Stagnant education. *Noncompetitive education* with the graduates and the production of knowledge.
- Low investment in RD&I *research* and development: *deprived teachers' profiles*; *insufficient* educational infrastructure.
- Poor *governability productivity* and *competitiveness* at all levels, low credibility of *teachers* and *researchers.*

2.3. Consequences of the Problems

The educational sector in Colombia has suffered from protracted failures in productivity and competitiveness, with large repercussions in the PSEd. In summary the consequences are:

- *Poor performance* of the students in international tests as PISA (The Programme for International Student Assessment) and TIMSS (the Trends in International Mathematics and Science Study).
- Very *low national competitiveness*
- Low *budgets* & academic sustainability; *high* costs for Government + *families* + HEI; and *high poverty* and digital gaps.
- Low investment in RD & I; Poor educational infrastructure, labs, libraries and pedagogic spaces.
- High *unemployment* in ICT.

- Bad isolated solutions. Deficiency *using* ICT *high leading technologies* as AI, KE, *architectures*, computational and mathematical models. No space for creativity, disruption and innovation.

2.4. Means for Significant Achievement

The means to achieve the goals are:

- *Mature student permanence policies*. OC. *Policy research*. Articulation HE-ME.
- *Edge technology management* for ME-HE institutions, for *deficient quality information systems* and for *absence of enterprise architectures.*
- *Productivity improvement of teachers and students*. Optimizing restricted resource uses. *Education internationalization.*
- Government policies aimed at addressing structural unemployment in ICT at all levels.
- Alternatives for students who cannot make careers of several terms.

2.5. Final Purposes. Competitiveness and High Quality of Education

The goals to achieve are:

- Continuous improvement and governance. Improved capacity for generating knowledge and innovation.
- High quality, economic and financial sustainability; national capacity to generate knowledge & technological innovation.

- Declining digital divide and supply the deficit on ICT talent. Massive *social innovation*.
- Compliance with social goals. Technology transfer to other HEI at marginal cost; technology transfer and appropriation.

2.6. THE END-MEANS MATRIX (EMM)

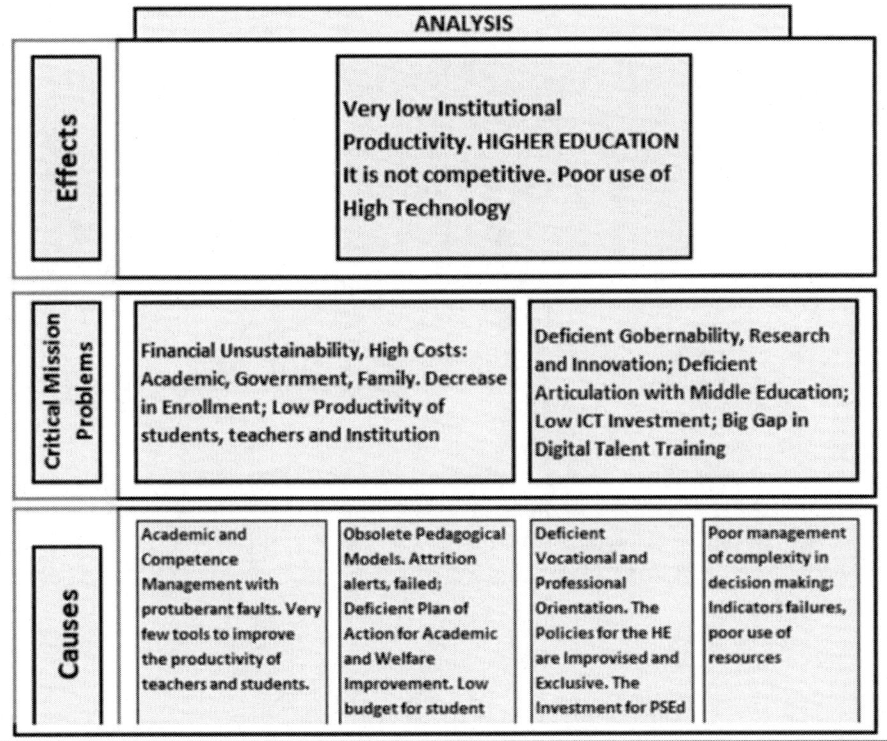

Figure 2.1. End-means matrix analysis.

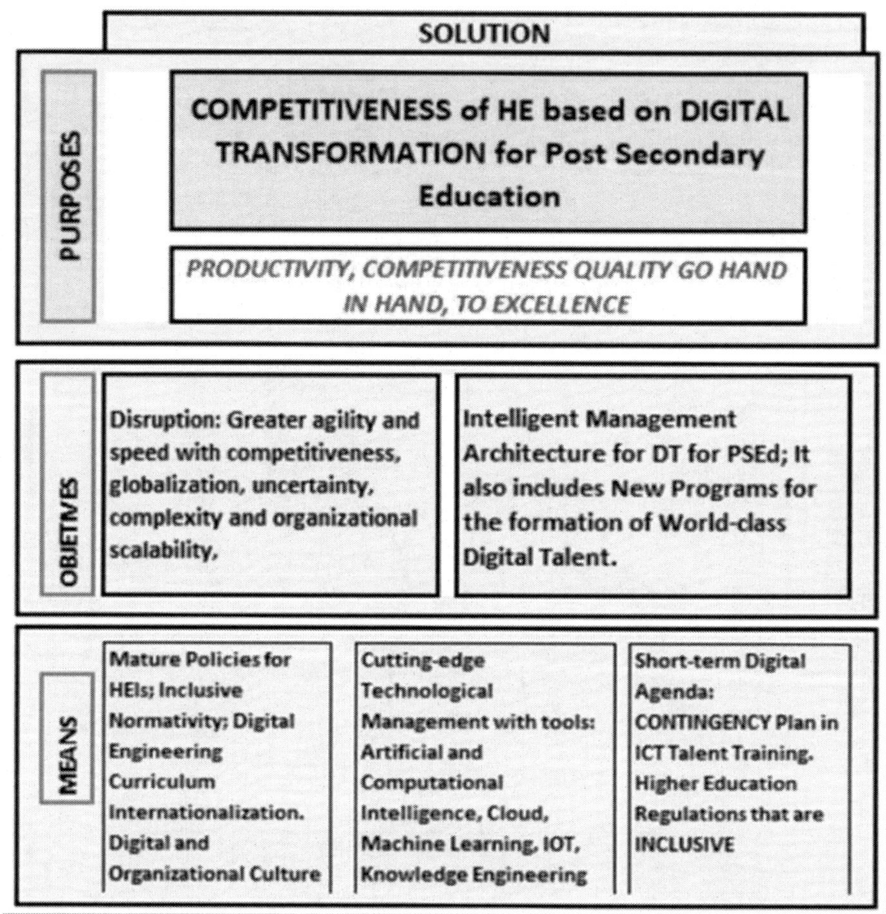

Figure 2.2. End-means matrix synthesis.

2.7. Productivity

Inventions supported by creative innovation improved productivity over the centuries by decreasing production costs with incredible solutions to solve problems.

The functions of production made less work than the contribution of capital and technology which increasingly require more skills, attitudes, ability to adapt to change (constant and not sporadic) and skills. It does not surprise us that the Capital (K) was substituting with the passage of time the human force, Labor (L) by systems characterized by the business organization.

In the third millennium, technical progress has been very important and increasingly fast coupled with communications supported networks integrating people who are not even in a location and empathize with what happens around: it is a multivariate change and costs a lot to identify the aspects of each other to isolate them and act on each of them. With 2 important rewards being on the crest of the wave (of the growth) as long as possible and solve socio-economic problems by meeting planned goals.

The above is aggravated to the extent that HE is *structurally* weak and *very* few of our educational *institutions are referenced in the world context.*

The Law of Tertiary Education in this country fractured the HE by creating 2 highly discriminatory aspects: the *professional* level and the T&T levels (professional technical & technological); one of apparently "good quality", and another, the "poor relative", in a giant gap of inequity undervaluing the T&T training from the social and labor context which claims an education of HQ.

Productivity and competitiveness is not everything, but it is almost everything: 100%.

Suffice it to mention a case of *investment* according to OECD (2016) per student in Colombia: 3165 USD. The countries of the OECD on average are at more than 3 times that value, and in the USA, it's five times higher.

It is foreseeable that an era is coming when productivity will have a great global acceleration (for the IV Industrial Revolution) and although it is paradoxical in this transit, electric power, TV, telephone and vehicles are still gaining the hand to the applications of the Digital era and of the socio-technical progress of the cyber-physical systems, and the consequences in productiveness.

Expected changes to digitization would have greater acceptability, but: When they really come to the market to have the profound changes in the *education*, industry and society in general?

What do we need to enter the new era to leave behind: *poverty*, the *exclusion*, the *inequality*, the *protuberant digital gap*?

2.8. Modelling for Digital Transformation

We have proposed to offer to the academic and productive community our conceptions, constructs and implementations redefined by propaedeutic cycles of HE, that we have been working for some time in our University, which is FESSANJOSE.

In the following Figure 2.3 we present a schematic synthesis of our aspirations in DT, that we are specifying with this publication, to make viable the entrance to the IV Industrial Revolution.

In the Table 1.3 of Section 1.2.1.12, Tools for DT built at FESSANJOSE, we present a list of our main constructs and publications around the topic of DT. Our contribution with a DT dual model intends to respond to the previous approaches, with successful and leading technologies in industry and commerce within the new digital economy based on knowledge and intelligent cogni-facts.

KE in education (KEE) (joint with big data and IOT, machine learning, and related), which is considered a part of engineering in the KS, deals with the organization, representation, use and application of education knowledge to design, build and deploy intelligent agents and expert education systems. The KEE is the top step of the engineering systems and the engineering of AI. They are our own research experiences.

2.9. MIR MATRIX

MIR : *MATRIX OF OBJECTIVES, IMPACTS, Indices,Results: Project - DT-UNIVERSITY DIGITAL TRANSFORMATION ARCHITECTURE*		
GENERAL Objetive	**Main Products and Outcomes**	**Indicator**
To build an intelligent MANAGEMENT ARCHITECTURE for the DIGITAL TRANSFORMATION of Postsecondary Education	High Technology Management Model for the Digital Transformation of PSEd	Number of teachers and administrative certified in DT
	Management System of skills, knowledge, scenarios and Smart Strategies	Digital Software Assets, Registered in the DNDA
	Enhanced Digital Engineering Training and state-of-the-art technologies	Number of certifications
Specífic Objetives	Plan of Action and Road Map of DT and IVth Industrial Revolution	Record Document
	Plan of indicators and indices-numbers for tracking and monitoring	Manual
a) Implementing DT tools based on innovation and re-architecturization of successful experiences	Digital Intelligent Tutorial System Engineering, Machine Learning and Artificial Intelligence and related.	Software, Patent and Manual
b) Intelligent Re-engineering of organizational and curricular processes of DT, based on innovation - export quality	Documentation Manual with IEEE standard, User Management Guides	Manual
c) Intelligent re-architecturization of the functionalities of Data Architectures and online tutorials. d) Incorporation of INTELLIGENT STRATEGIES, WITH METACONOCIMIENTO AND 2 LEVELS OF REASONING. Use of the mathematics of artificial intelligence. Fuzzy logic for knowledge processing	Models of: 1- Gobernability of the DT, 2- Digital Social Projection. 3- Institutional Self-evaluation. 4- Classification of Institutions in DT 5- High Quality PSEd. Continuous improvement TOWARDS EXCELLENCE. 6- Model Methods and Methodologies of DT	No. of Digital Software Assets, Registered in the DNDA and Colciencias

Figure 2.3. (Continued).

e) Continuous Improvement and Academic and Administrative Governance. Preparation of the MANAGEMENT CAPACITY of CHANGE	**IMPACT OF PROJECT**	**Coverage and Scope of the Project**
f) New Programs in Education of PostSecondary, by Propedeutic Cycles, for the formation of Digital Talent towards the IVth Industrial Revolution	Social Appropriation of Digital Knowledge: 1 Book and 3 International Papers to defend in world-class events with global diffusion	**Local and National Level**
g) Greater Mobility related to COMPETITIVENESS, globalization, uncertainty, complexity and organizational scalability: Technological Breakdown	Scientific Community Empowering: at least 100 certified teachers in Digital Technology Management	
	Generation of Digital Knowledge: Tools and Intelligent Constructs Based on Digital Knowledge.	

Figure 2.3. MIR: Matrix of objectives, impacts, indices, and results.

Conclusion of the Chapter

The sociotechnical progress of the countries is explained by the high quality of *education and scientific research.* Reason for which the DT, within the space of the ICT, is a very good strategic exit for the countries, especially those that are in the way of the progress.

In countries like Colombia the social sectors are increasingly hardly hit by the weakness of the economy, the unequal distribution of income, the armed violence, and political conflicts, in addition to drug trafficking.

The only way out is a solid education at all levels, HQ, which requires emulating more than proportionally, the productivity of the country, and the social, industrial, and governmental sectors and all its agents. University teachers play a great role in this commitment, preparing young people to compete in the global society, and doing research to promote the national capacity to generate knowledge.

The use of MIR and EMM matrices has facilitated us to synthesize the knowledge of the PSEd sector, and also to infer the requirements and objectives to achieve with the DT.

Chapter 3

POSTSECONDARY EDUCATION DIGITAL TRANSFORMATION

ABSTRACT

The *Dual Model of Digital Transformation for Higher Education*: on the road to the IV Industrial Revolution, institutions focus on the massive and optimal use of ICT in different organizational levels including human-digital talent, business structures, processes, inputs, products, services, and obviously the business model. The DT of the *university* necessarily implies *the cognimatics* (information technology for the Knowledge Society) of all the institutions and their social insertion in this country with new models: institutional and educational. The duality is related to curricular digitization and institutional digitization, that is, knowledge and data. The pillars of the architecture are *research*, *entrepreneurship*, *financing* and *social projection*, recognizing from the beginning that knowledge has its ethos in the *university*. We present in this chapter:

1 The productive *ecosystem* of the DT.
2 The DT that enhances the ability of the university to generate knowledge and innovation towards the empowerment of intelligent digital capabilities.
3 A new economy that requires social transformation and technical-scientific progress, seeking substantial improvements in student,

teaching and institutional productivity towards high quality and competitiveness.

4 New DT human talent required by the new industry 4.0 of knowledge and intelligence. The hyper-personalization of the student is demanded by competences and abilities. We have developed a methodology for DT that responds to this complexity, applying some intelligent constructs that we have published in IEEE events during the last 10 years, leading the DT in *postsecondary education* that covers the *professional-technician*, *technological* and *professional* levels.

Keywords: digital transformation, postsecondary education, digital talent, ICT, artificial intelligence, fuzzy dashboard matrix, DT ecosystem, DT methodology, DT sorting

3.1. University Digital Transformation

On the way to the imminent IV Industrial Revolution, the companies of all kinds are focused on the massive and optimal use of ICT in different organizational plans including human talent, organizational structures, processes, inputs, products, services and obviously the business model. Nowadays, 70% of the top companies have robust DT teams, and 40% of them are being driven by high ICT and artificial and computational intelligence.

On the other hand, very few Colombian universities have tackled the problem of competitiveness in HE, and much less have made the transition to the IV Industrial Revolution by taking the leap, thus, receiving a tremendous blow due to academic obsolescence.

The universities should play an important role in the reduction of the digital divide, and even more, they must design new programs that face the formation of the digital talent that is required by this new paradigm; or we will be late again to change.

The gap represented in Figure 3.1 is the difference between today and the future, mainly in terms of:

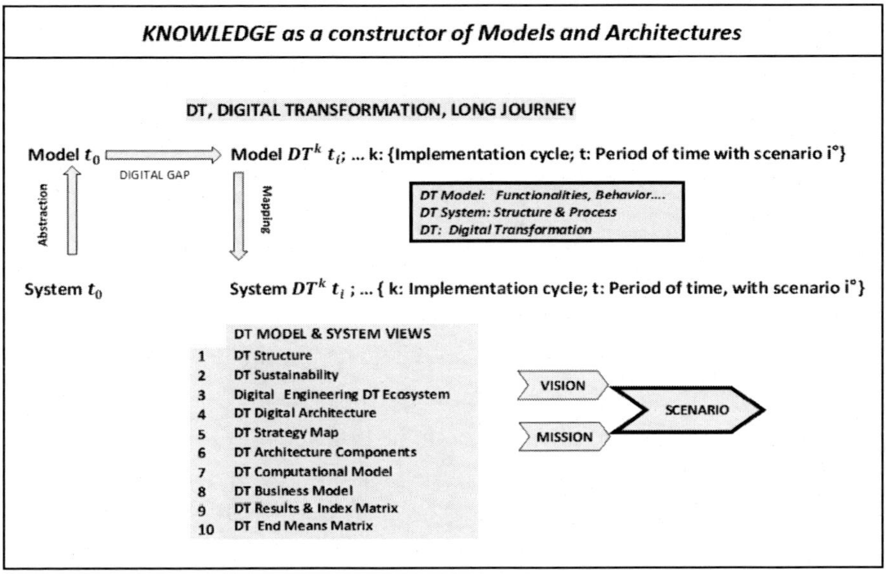

Figure 3.1. Digital GAP.

- Technology
- Digital culture and human talent
- University organization
- Organizational model
- Educational model
- Productivity
- Competitiveness
- High quality
- Institutional performance.

The engine is the *knowledge,* as it's also the *constructor* of *models* and *architectures* that evolve with the DT, which is not a goal of arrival, but a station of passage, in this long journey towards progress.

3.2. TRANSFORMATION INSIDE PSED

DT seeks to significantly improve the *productivity*, *competitiveness*, and *quality* of PSEd institutions. DT is an accelerated evolution. It is also *revolution* because of its radical and structural implications for people as for infrastructure that also demand new educational and business models. It requires accepting that the potential of DT is so aggressive: there is no time for an institution to take the position of "wait to see what will happen"; in fact, e.g., you can still work with outdated technologies, but it does not means getting rid of all the systems to buy them from new. *Digital competences* are no longer the plus for an institution: *they are an imperative*. It is the set of DT strategies what constitutes the governance of the digital activities of its components, driving the excellence in digital operationalization.

The **visioning** process is based on the assumption that figures of the future lead actors' present behaviors, guide choices, and influence decisions. Vision comprises players, stakeholders, values, wishes, and behaviors. Vision is a framework of the DT future shared by an ICT community; and that vision is adaptable into reality.

Our *DT Model* is presented with several visions, namely:

1. The **Ecosystem** of the DT.
2. The **Structural** vision, or the pillars of the *university* DT: *research, entrepreneurship, digital talent, budget* and especially the *social projection* of the *institution*.
3. **Strategic Map** of the DT showing scenarios, actors and vision-mission.
4. **The architectural components for DT**: the eight architectures developed and implemented for our DT Model applying some intelligent constructs that we have developed and documented in the last 10 years at FESSANJOSE, leading the DT in PSEd.
5. **Digital 360° architecture** of DT: Academy and Administration LOCUS: This architecture is the *digital portfolio* and implies the organization of the subsystems to obtain better and/or new functionalities based on knowledge to obtain an intelligent behavior. The architecture-multilayer approach is a system of systems (SoS), which ensures compliance with government policies, norms and standards, in a highly complex social institution with digital assets; this approach describes the subsystems at a higher level, where a system is composed and with the

protocols by which they communicate. It provides a 360° business vision map and a planning framework for commercial and technological changes.

6. **The computational-mathematical perspective of DT**, identifying endogenous, exogenous variables and interrelations.
7. The synthesis: **Matrix of End-Means** (EMM) that summarizes in DT: Where we are? And where can we go?
8. **The MIR Matrix**: DT *objectives*, *impacts-indicators* and *results* to be obtained with the DT.
9. The **Dynamic Model of the DT**: system based on *computational intelligence*, representing the control system of all its components to achieve the completion of DT. It is an iMIS, *intelligent management information system* for PSEd that shows the dynamics of DT control integrating several multilevel system hybrid architectures, as a space to respond to the solution of HEI problems, tending to their *competitiveness*, specifically pointing out the way these modern technologies can be included for their adaptation and evolution in PSEd in post-modernity, making *governability*, *teaching* and *student productivity* compatible with educational high quality, the purpose of DT.
10. The **Interface of the Results of the iMIS** such as: *high quality metrics, indicators* and *values* of *management*, *desertion*, *answers* and *plans*. These results can be decisions, information (e.g., *permanence*), knowledge required, solutions, inferred plans, policies and objects of *continuous improvement* for the sustainability of the transformation and quality among others.
11. The **Input Interface**, with such as: *data*, information and knowledge acquisition: where the attributes, parameters, values, problems (e.g., desertion) are registered in such a way that actively and dynamically obtain and structure. The *problem* is expressed as the type of decision to be made, e.g., the need for information in the complexity of the *student dropout*: i.e., for students trained in mathematical competences, a prospecting of system variables, as the decrease rate in the *excellence - current state* gap

3.2.1. Raison d'Être

The rationale of the DT is synthesized below:

1. Unprecedented *prospects* befit the paradigm changes, given that they promote the use of high technology that contributes to the development improvement of new and current educational products.
2. The fulfillment and scope of the requirements demanded by the DT (see next section) *increases the know-how* with the escalation in the

institutional capacity to generate knowledge, as it concerns a PSEd institution.

3. It can be shown that DT *optimizes* relations with *users*, given that the client (specially the millennian) is the center of the innovations that happen within the DT.
4. Institutional solutions with accessible technologies are evidencing reductions in operational costs and improvements in productivity, competitiveness and high quality in academic-administrative sectors.

3.2.2. What Is the DT Requiring?

DT is not free lunch. Very important efforts are required for its design and implementation in the different scenarios or the locus of the PSEd: *technology* (drivers), *organization, users* (teachers, students, parents, infrastructure, planning, government) and a new organizational culture with new digital talent.

3.2.3. Functional Requirements

The Model evaluates fundamental aspects that are important for a universal DT of the *institution*, such as support and executive commitment with the digital strategy, recruitment and provision of digital talent, and *digital marketing* focused i.e., in the specific capabilities of a company's digital marketing functionalities, such as how DT is compatible with the brand strategy based on the digital business. How does digital support interact with sales and services, including contact point of integration and technological sophistication?

DT is necessary for reengineering the administrative processes with support from software intelligent architectures, towards the roboticzation of processes and their autonomous performance and decision making.

3.2.4. Non-Functional Requirements

It is required to make the decision *now!* otherwise, competitors will take the lead in the market and threatens our business survival. Budgetary and investment resources must be so flexible as to hire consultants, educate, train, and maintain the technological infrastructure at a good level of performance.

DT also demands globalization and digital curricular and administrative innovation towards a flatter organization (i.e., under coupling), which is why new curricular, educational and business models are imperative.

To understand that DT is not fashion or simple novelty, important institutional efforts are required on how to approach cultural aspects, organizational challenges, technical challenges and incessant informational and cognitive challenges.

Also, it needs to address the problem of cybersecurity, privacy and data protection, which also requires new confidentiality clauses for officials.

3.2.5. High Quality PSEd

The HE needs to be integrated in its 3 natural levels: *professional technician*, *technological* and *professional*. It must feed on the DT to improve more than proportionally its productivity (institutional, for students and for teachers) as an unavoidable step to competitiveness, and there, to the high quality.

Security requirements require reasoning about the dangers of innovations and its associated risks and/or side effects. The threats and weaknesses identified in the strategic plans must be quantified, and insurance must be taken when the previous cannot be easily taken from the *digital talent requirements*.

A DT human resource with disruptive thinking requires continuous training throughout the entire organization with permanent evaluation of its skills and commitments.

Finally, DT needs to mobilize new experts with the conviction that digital competences are no longer the plus: *they are imperative!* i.e., digital skills and abilities are not short-term flashes of light.

3.3. DT Conceptual Foundations

The main challenges are the reduction of profit margins and destructive unfair competition. The adoption of our model supported by the analytics and the DT itself in general, permits the performance with speed, agility, re-mobility, personalization, revolutionizing the digital experience of the client, the intelligent re-engineering of the academic and institutional processes, and fully understand the needs and behavior of the client. All this will be our central objective to occupy a privileged place in *global competitiveness*.

The same automation is an opportunity to re-imagine the organizational processes, with a new educational-university and business model. This presents us with great possibilities to take advantage of our automation platforms and our ecosystem (defined in section 4.4) to bring our students into this ethos of knowledge with a constant innovation in science and in social area.

In the new world that enables DT, e.g., with exponential technologies, some universities will survive, and others will not. The institutions that perform agile and establishing a digital culture of change will participate in a wide spectrum of clients-students and of companies that require consultancy.

It is a changing landscape as regards to how digital is changing the institutional and governability rules.

3.3.1. Previous Concept of 'Transformation', or 'the Transformed'

In engineering, we learnt about transformation. Next section we discus about it.

3.3.1.1. The Term Transformation

It refers to the action or procedure by which a system is modified, decomposed or changed, in a way that maintains its identity. It has as associated manifestations: transformed, projected, evolved, from one body or component to another organism. In mathematics we have as integral transforms the corresponding, v.gr., Fourier transform, Hilbert transform, Laplace transform, Wavelet transform, and discrete transforms such as Z transform, among others. They have different uses and applications, for example, the Laplace transform is frequently used for the resolution of ordinary differential equations, facilitating the use of integral-differential equations by algebraic manipulations (time domains versus frequency domains.) Then we can discuss social transformation (that redirects to a "social change"), political transformation as a projection of the environment, as well as a transformation of nature as a result of the evolution or interaction of living groups with their environment.

In regard to the past, struggles for productivity, and very complicated competitiveness: projecting ourselves on the digital space makes it easier and less expensive to achieve competitiveness and quality with important improvements in productivity.

3.3.1.2. Definition of Digital Engineering

The essence of *the digital* is that it is value oriented! Today, users want solutions that require a consultative approach, combining the various existing assets of the client. Users already host a large amount of data, information and knowledge, albeit in silos, which must all be interconnected to obtain ideas that build value.

The digital has changed the procurement behavior. However, there is a growing awareness about that companies cannot offer a superior digital experience without making fundamental changes at their back-office. Today, it is not about having an application, but about quickly fulfilling a promise.

Digitization is the way to go and having a relevant strategy that provides mobility, speed, effectiveness, and proactivity, i.e., all the essential characteristics for progress: it is an accelerated transformation, which is why

DT refers to organizational, cultural, technological, visionary and perceptions changes, maintaining identity as a system. It is the confluence of mega-trends (which change the way of doing business and their operation, and the way to see in real time what happens inside and outside the organization, based on data, and far from simple intuition or subjective considerations), so we affirm about the DT:

- It does not depend on the size or class of the company, with the exception of universities which have an additional dimension, more than one enterprise: *the academic* arista.
- It is not a simple challenge.
- It is not a so-called new revolution.
- It is not about absorption of technology.
- It is not a new paradigm.
- It is not isolated and unrelated automation or computerization.

Even, on the contrary, DT:

- Is a transformation of an era for the design and construction of the future of a system, which due to its radical implications for individuals, such as organization and infrastructure, demands new institutional and business models, and the redefinition of globalization and economic, social, academic, industrial and governmental practices.
- Is an organizational imperative to compete and survive successfully in the near future.
- Is the creative re-invention of the current institution.

These conclusions have required a lot of research during the last half century, with substantial innovation from a new *digital engineering* (DE).

3.3.1.3. DT Has Several Aristae

These are multiple edges or dimensions that characterize the DT, namely:

- Business and academic models, including the educational one, because the new organization is demanding them.
- External and internal user experiences to be used.
- Organizational processes with robotization and AI.
- New digital and culturized human talent that needs to be more versatile, and access greater variations in communication formats, such as the *omni-channel* way with users and customers, that constitute the imperative for the continued development of digital skills for all people, especially those who are in productive ages.

3.3.1.4. Difficulties with the DT

There are ambiguities related DT, i.e., to pass from '0 paper generation' to 'the application of DE in all aspects of the socio-economy'. Consequently:

- Several DT models have been defined irregularly, they have been applied in a dissimilar way, and the measures to address them have tended to be unreliable and difficult to compare.
- Resistance to change: there is no consensus, and there is much lack of commitment; also, there is resistance to the possible side effects or dangers involved in the change.
- Rigidity of corporate culture and attachment with obsolete technologies of the last century are observed, but in turn, this does not mean getting rid of all the legacy systems and have to buy them again.
- Also, absence of digital strategies, the *vision*, the *mission* and the values duly updated with a focus on the user.
- Missing alignment synchronizing corporate strategies with the new digital strategy.

Metaphorically, DT seems to be a tsunami or great earthquake for the company today in the global business models. The traditional models are being completely broken. These major changes alter the course of each economic sector, including IT services. Visionary companies can visualize emerging opportunities and transform themselves to exploit these opportunities.

3.3.1.5. Digital Capabilities

Unfortunately, this attention on emerging opportunities has led to confusion in the market about what does mean this "digital institution" transformation. Conceptually a triumphant transformation requires the development of digital capabilities, which means:

- *'Development'* is a process by which people, organizations and society are separated from the obsolete, and strengthen, create, adapt and maintain digital capacity over time. The *development* of digital capacity is different from the *"construction"* of digital capacity. The development of digital capacity is an internal process at the institution, driven by stakeholders, or also by stakeholders that can be found outside the country. Capacity development refers mainly to the *responsibility* of suppliers or customers and stakeholders, so they play an important role.
- Understanding the challenges of capacity building and the very development of digital capabilities in institutions. There are as many differences as there are similarities in how digital capacity can be built.
- Enlist/provide expert advisors/mentors on sustainable digital capacity requirements to meet the specific needs of each institution, which is a non-trivial task. The objective is to provide expert advice on sustainable capacity requirements to meet specific needs that will improve the quality towards competitiveness.
- Among the capabilities mentioned is the digital commercial agility. This form of agility is fundamental to allow organizations respond

to the high-speed and unpredictable change characteristic of the digital irruption.

It should be noted that business transformation is digital when it is based on an information technology infrastructure. This focus on digital technology is different from other possible factors of transformation, such as political, social, cultural or based on economic changes; and it is even more so, as paradigm changes. It is also worth mentioning that in technologies and businesses, the models that sustain the DT are not fixed, i.e., they are not of uniform use or application because of their variation over time and also, to some extent, because their application in industrial or commercial sectors.

It is clear that the technologies are most significantly associated with the transformation of the digital business:

- Extreme tools and analysis applications, including "big data", machine learning, IOT (including connected devices and "smart" networks).
- Tools and mobile applications.
- Platforms on which to build compatible digital capabilities, such as cloud solutions and related, and application markets.
- Tools and applications of social networks.
- AI and KN.

3.3.1.6. Implications of the DT

The architecture of DT extends the benefits of digitization to the entire organization and reduces the human talent gap that DT requires.

Other implication is to see the DT as "the acceleration of commercial activities, intelligence of academic and organizational processes, skills/capabilities, proficiencies management and mathematical-architectural-computational models, to make the most of the changes and opportunities junctures of digital technologies and their impact in a strategic-disciplined-rigorous and prioritized way".

But more than just acceleration, the DT is seen as the institutional requirement to overcome the digital irruption and stay competitive in a digital business ecosystem in rapid development and evolution. It represents a paradigmatic change in the institution experiences/practices and productive strategies as a direct response to opportunities-conjunctures given by technological evolution.

3.3.1.7. Opportunities for DT

The World Economic Forum (WEF[12]) states: "They used to take Fortune 500 companies an average of 20 years to reach a valuation of one billion dollars, today's digital start-ups are coming in four years."

While increasing competition is a powerful motivator to include the DT, so is the income opportunities. The WEF estimates that "the combined value, for society and industry of DT in all industries is over $ 100 billion over the next 10 years."

The opportunities in relation to the effectiveness, the learning of the experiences of clients and the decisively involving the employees for more than proportional achievements, stands out for:

- Operational efficiency: How can digital opportunities be exploited to provide cost-effective services or products of high quality?
- User commitment: How can you take advantage of technology to reach existing and potential customers in more meaningful ways? This is especially relevant when many people these days are "connected".
- Employee engagement: How do you use technology to retain and cultivate new talent? How to promote "a happy employee, who is a committed collaborator and who remains committed, innovates and contributes to the organization?".

[12] https://www.weforum.org

3.3.1.8. How to Handle the Digital Opportunities?

The value of each identified digital opportunity must be quantified, both internally and externally, by calculating its expected values. Assigning the digital junctures to the objectives of the institution is a great place to start. This helps to maintain the strategic direction (mission-vision) and eliminate any idea that is not a plus as a value.

It is necessary to go one-step further: to determine exactly what is relevant for the institution. For this, organizations other than nonprofit making, or from the industrial or commercial sector, may be looking to grow in:

- Attract and retain parents within the educational network.
- Sales and conversion rates.
- Knowledge of the student's stratum and the expectations of their parents.
- Value for stakeholders and others as education authorities.
- Optimize the rotation of inventories and talent staff.
- Appropriately manage ICT costs, e.g., use of *the cloud*.

In addition, one should reflect on the current possibilities of information, technology and digital talent, and estimate the costs of carrying out the operationalization of the DT model. The fuzzy-morphological-matrix analysis is very useful.

Obviously, about the risks, as in any large project: management of associated risks and conflicts is a must.

3.4. PSED ECOSYSTEM OF THE DT

3.4.1. Definition

The *PSEd ecosystem* is defined as a network of HEI agents, accredited or not accredited, that are articulated around the DT from the needs of the country that respond to the PSEd contexts' characteristics in order to increase the student, educational, academic and institutional productivities, to finally make the transfer of knowledge in function of the substantial improvement of the *HE competitiveness* that integrates all the PSEd's levels without exclusions, for the social and productive development of the country.

Also, it is composed of research programs, experimental development, technological development and/or innovation that encourage the generation, transfer and appropriation of knowledge in collaborative work with international and national peer entities.

The digital ecosystem is understood as the set of benefits and requirements of diverse nature that are provided from and through telecommunications networks, the set of infrastructures and associated services that enable the provision of such services, and the interaction between the providers of services of a different nature that constitute the extended value chain of Internet services.

As the set of infrastructures and means associated with the provision of content and services through the Internet, the *ecosystem* constitutes a new a subject of fundamental analysis from which public policies must be considered for definition, in areas as diverse as the digitalization of productive processes and the protection of users' privacy, which has already begun to be addressed recently by various entities and international organizations.

The concept of digital ecosystem defines a new industrial context and economic and social impact resulting from the massive adoption of digital technologies.

The study of the digital ecosystem involves three *loci* or dimensions: (1) new modes of information production, knowledge and content, (2) different social behaviors related to the use and consumption of goods, and (3) a more important economic and social impact than information and communication technologies considered in isolation.

3.4.2. Ecosystem Components

According to the scope of the DT in terms of competitiveness, the PSEd ecosystem must be consistent with the academic and administrative processes with ***international*** reach of its new models (institutional, curricular, pedagogical and educational), as is appreciated in the Figure 3.2 bellow.

- Structurally the components are:
  1. Locus of culture and human resources: a university's approach to digitally driven innovation and how it empowers employees (teachers, staff) and students with digital technology.
  2. Organizational Locus: how a university is aligned to support the digital strategy, governance and execution.
  3. Technological Locus: the use and adoption of a new technology that is emerging, and a new infrastructure and platforms.
- Functionally the DT governability is intended on different layers:
  1. Empowered users, with their perspectives on DT,
  2. Application layer,
  3. Layer of services on Social Projection,
  4. In addition, the infrastructure layer.
- Coherently, it's understood that the ecosystem architecture for the PSEd should guide towards dual productivity (institutional and academic), competitiveness and high quality.

3.4.3. Advantages

The ecosystem for the PSEd's DT contributes to the improvement of the HEIs quality through research, experimental development, technological development and/or innovation, with solutions to the needs of the country.

3.4.3.1. In the Short-Term

a. Consolidates the capacities of research, experimental development, technological development and/or innovation.
b. Contributes to the improvement of the formation of high level human talent that builds the PSEd to compete in the global society, pave the way to the IV Industrial Revolution, and survive successfully in the Knowledge Society.
c. Improves the visibility and positioning of the *institution* at an international level.
d. Contributes to the improvement of institutional productivity through the intelligence of the organizational and academic processes.

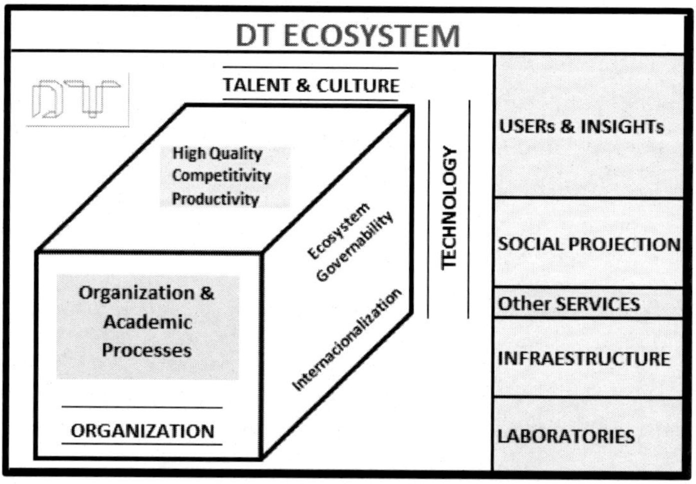

Figure 3.2. DT ecosystem: Scenarios, components and dimensions.

3.4.3.2. In the Long-Term

a. Contributes to the improvement of productivity and competitiveness of the country through the generation of a portfolio of products and services, e.g., research and innovation processes.

b. Collaborates with the cognimatics and digital articulation/ coherence between national and international HEI with the other productive and social sectors of the Colombian economy.

3.4.4. PSEd Ecosystem Locus

In the dynamics of the DT's performance, the *loci* are: the *organization*, the *culture*, the *digital talent*, and the *technology*.

3.4.4.1. Culturally

It is the digitally driven innovation and how to empower employees with digital technology. The competitive strategy depends on digital technology. The directives and top executives support the digital strategy. There is: an investment in targeted digital education and training at all levels of the organization; a clear communication of the digital vision both internally and externally; and taking of measured risks to allow innovation.

3.4.4.2. Organizationally

Certain concerns must be answered, including: How aligned is it to support digital strategy, governance and execution? Does the organization prioritize user experiences on functional silos (data storage towers)? Does it dedicate the required resources? Is the organizational model collaborative? Do the suppliers offer value that improves our digital skills? Do we have clear and quantifiable objectives to measure the success of our digital strategy? Do we measure the functionality of the channels together, to achieve the desired result? Does the user's perspective guide the digital strategy experientially? The user experience allows us to infer about elements of design and digital development.

3.4.4.3. Technologically

Is our budget for technology practical to allow flexibility of priorities and strategies? Do we take advantage of modern architectures (API, cloud, big data, IOT, IA, robotization of processes, etc.) to promote speed and flexibility? Do we use digital tools to promote innovation, collaboration and employee mobility?

3.4.5. Constituents

The following table shows the different components of the DT Ecosystems.

Table 3.1. DT ecosystem components

INSIGHT

USERS	APPLICATIONS	SERVICES	INFRAESTRUCTURE
TEACHERS	TEACHING	SOCIAL PROJECTION	ICT-DIGITAL
PARENTS GOVERNMENT INDUSTRY	RESEARCH EXTENSION		UNIVERSITY PHYSICAL
STUDENTS			

3.5. DT Structural Model

The DT in a HEI that has its ethos in a complex system, has numerous determinations and edges of problems.

We presented in Section 4.1 different architectural views for the DT.

3.5.1. Functional Architecture of the University DT

The architecture has dual functional aspects: academy and organization, of the *intelligent educational architecture* in terms of its pillars and sustainability components.

According to the previous definitions, the DT is located in the space of the institutions' productivities, the agents, the organizational processes, and the users, equally, according to the *ecosystem*.

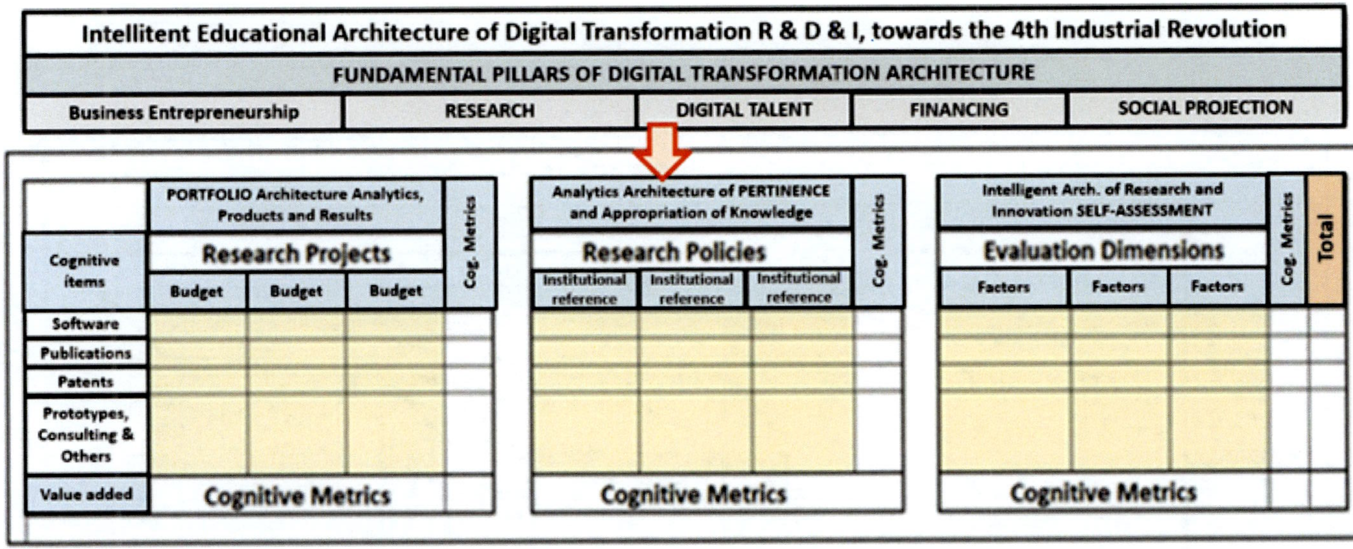

Figure 3.3. Pillars of the DT model.

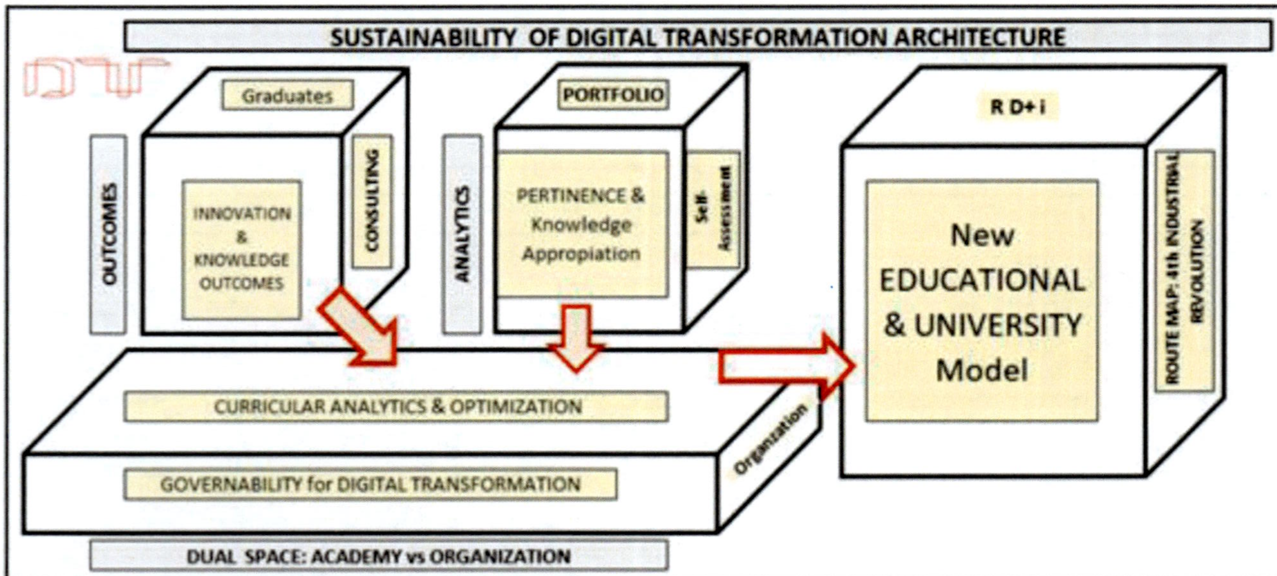

Figure 3.4. Sustainability of DT architecture.

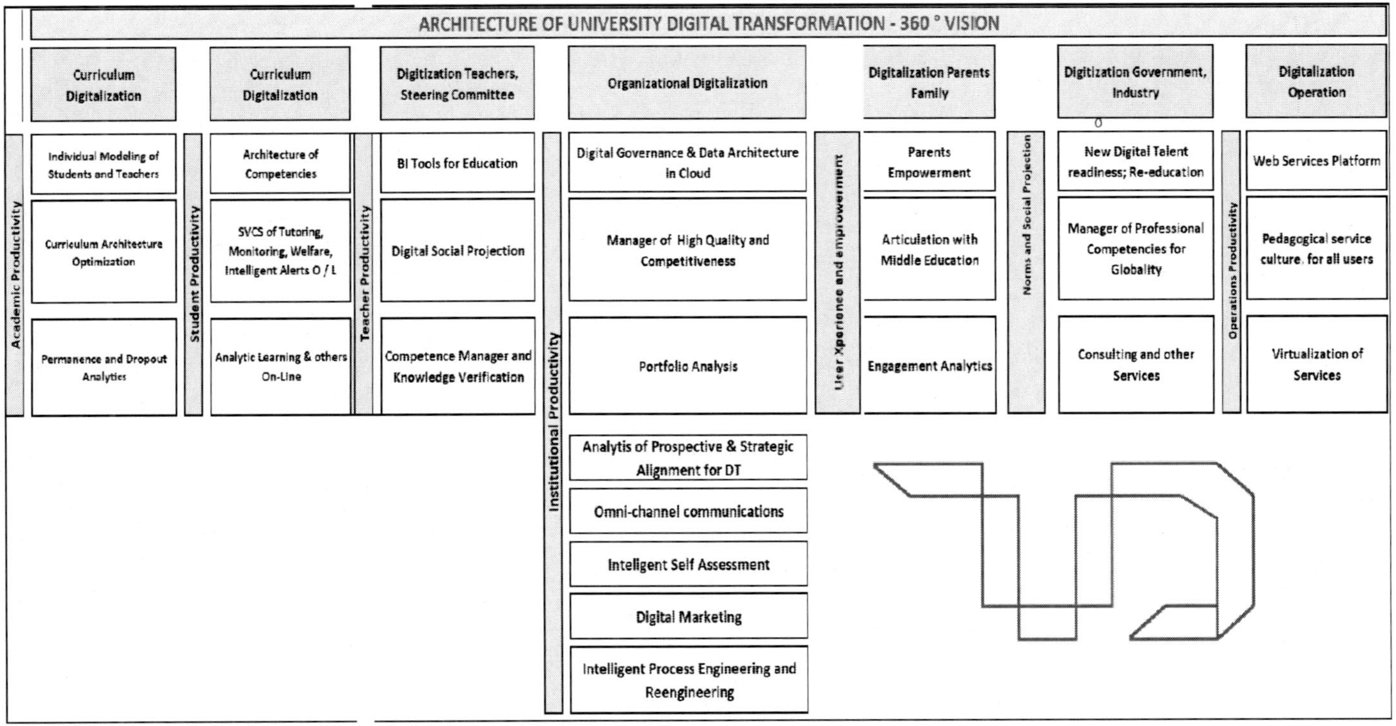

Figure 3.5. DT 360° digital architecture.

In the case of the PSEd, the DT has 2-dimension spaces:

- *the curricular and*
- *the administrative.*

The agents or actors are: the students, the teachers and the staff. The stakeholders are: the parents, the CIO of the institution (superior council), and the government, and related.

The fundamental pillars of the DT Model are:

- The enterprise, taking into account that the DT promotes the re-invention of the institution.
- The research, combined with innovation, for the generation of *intellectual-digital assets*.
- Digital talent, endowed with digital expertise to promote, implement and maintain the system towards the IV Industrial Revolution.
- The required budget, given the effort needed to get rid of the past and adopt new and extreme technologie.
- The social projection of the institution both inside and outside, for the transfer of knowledge and technologies that promote the productivity and competitiveness of the country.

Model Architecture (Figure 3.3) is integrated by 3 essential components:

- Portfolio analysis of produced digital and intellectual assets.
- The analysis of the relevance of research policies and guidelines.
- The Self-Assessment evaluated in different relevant dimensions.

3.5.2. The Sustainability of the DT

The DT sustainability model is presented in Figure 3.4, where it is clearly shown the duality *academic space/administrative space*, whose characteristics are:

1. The generation of new knowledge translates into publications of scientific-technical papers, properly indexed books, patents, vegetable varieties, software, or new animal breed, etc.
2. The technical-scientific development and innovation is characterized by: prototypes, industrial designs, innovations, pilot plants, technological products, such as software certified by the DNDA[13] and COLCIENCIAS[14], business products, scientific-technical consulting, standards and clinical guidelines, etc.
3. Appropriation and social projection, which is related to:

Strategies and technologies of communication and diffusion of knowledge (events-publications), generation of digital contents, networks of specialized knowledge, seminars-workshops of creativity, etc.

3.5.3. 360° Vision of Digital Architecture Functionalities

For the institution, the DT Model, in a more than proportional way, improves the organizational productivity, optimizing *pari passu* the student permanence, minimizing the desertion, and satisfying the minimum commitments and the limitations of the different resources of the institution.

Digitization is the process of moving information from things or paper to digital forms (e.g., computer files, IOT, connected data elements representing a product managing a configuration and all its dependencies) to transform business, moving to a digital business.

3.5.4. Digitization Functionalities

Architecture implies the organization of subsystems to obtain better and/or new functionalities based on knowledge to obtain an intelligent behavior. The architecture-multilayer approach is a system of systems (SoS),

---

[13] National Directorate of Copyright, Colombia, http://derechodeautor.gov.co/home.

[14] Administrative Department of Science, Technology and Innovation, Colombia, www.colciencias.gov.co.

which ensures compliance with government policies, norms and standards, in a highly complex social institution with digital assets. This approach describes the subsystems at a higher level, where a system is composed and with the protocols by which they communicate. Provides a 360° business vision map and planning framework for commercial and technological changes.

The dual academy–organization digitization includes the following digitalization parts:

- Curricular, towards academic productivity
- Student, for student productivity, user experience
- Teachers and staff, i.e., for teacher productivity
- Organizational, for institutional productivity
- Parents, for the user experience
- Government and industry, with the new requirements
- The other organizational aspects
- And operational: looking for the operational productivity of the institution.

Figure 3.5 identifies most of the associated factors of interest of *digital architecture*.

Below, we describe succinctly each digital component.

3.5.5. Curricular Component, towards Academic Productivity

- Individual Modeling of Students and Teachers: styles of learning and teaching.
- Curriculum Architecture Optimization: using our tools of curriculum knowledge management.
- Permanence and Dropout Analytics: our implemented intelligent architecture for student desertion, both in the face-to-face and the and on-line methodologies.

3.5.6. Student Productivity, User Experience

- Architecture of the Competencies: knowledge and competence manager for productivity.
- On/line services of tutoring, monitoring, welfare, intelligent alerts, and others.
- Analytic Learning & other O/L pedagogic models for student productivity: Student Intervention (*expert system*).
- Digitization of access to bibliography according to the preferences and life project of each student.
- Digitization of the student's life cycle.

3.5.7. Teachers and Staff

- BI Tools for Education, i.e., for teacher productivity.
- Digital Social Projection: institutional extensions.
- Competence Manager and Verification of Knowledge for teacher productivity.
- Digitization of the university campus.

3.5.8. Organization for Institutional Productivity

- *Intelligent process engineering and reengineering*: new processes profiles.
- *Manager of high quality and competitiveness*: architecting the HE quality.
- *Portfolio analysis* of university results and research outcomes.
- *Analysis of prospective & strategic alignment for DT* in order to stablish and maintaining the *DT strategy.*
- *Omni-channel communications* with all users visualized in the DT ecosystems.

- *Intelligent self-assessment* by departments, faculties and overall.
- *Digital marketing* in order to improve the university income.
- *Digital governance & data architecture in cloud* using extreme technologies as a part of the IV Industrial Revolution.

3.5.9. Parents for the User Experience

- Parent empowerment in order to use their useful experiences.
- Articulation with ME: an imperative for HE integration.
- Permanence *analytics*: the student retaining is very crucial in the way to the HE quality.

3.5.10. Government and Industry with the New Requirements

- New *digital talent* readiness; re-education at all levels for the new ICT of the IV Industrial Revolution.
- Manager of Professional Competencies for Globalism: demanded universal PSEd curricula.
- Consulting and other services: *empowered* university social projection.

3.5.11. The Other Organizational Aspects

- Web Services Platform: for both institutional and academic purposes.
- Pedagogical service culture for all users: the DT culture.
- Virtualization of services: digitization in all of them.

3.6. Dynamic Vision for DT, the IMIS: USANJOSE-ARKE-INTELL

This iMIS integrates the architectures, components, and subsystems that make up the Control System of the entire DT, as shown in the Figure 3.6.

3.6.1. The iMIS is the Digital Transformation Information System for PSEd, Based on Computational Intelligence

This information system of the DT, shows us the dynamics of the control which integrates several hybrid architectures of multilevel system, as a space to respond to the solution of HE problems tending to *HE competitiveness*, pointing out specifically how these new technologies can be included for their adaptation and evolution at the *university*, in modernity, making compatible the *governance*, *teaching* and *student productivity* with the *educational quality* towards the *DT*.

The *results* of the iMIS are: high quality metrics, indicators and management values for desertion, responses and plans. These results can be: decisions, information (e.g., of *permanence*), required knowledge, solutions, inferred plans, policies and objects of *continuous improvement* for the sustainability of the transformation and quality, among others.

The input *interface* (data, information and knowledge acquisition): is where the attributes, parameters, values, problems (e.g., *desertion*) are registered and, actively and dynamically, obtain structure.

The *problem*: is expressed as the type of decision to be made, e.g., in the complexity of the student dropout and the need for information about it: i.e., for students trained in mathematical competences, a prospecting of system variables, e. g. decrease % in the *excellence vs. current* state gap, etc.

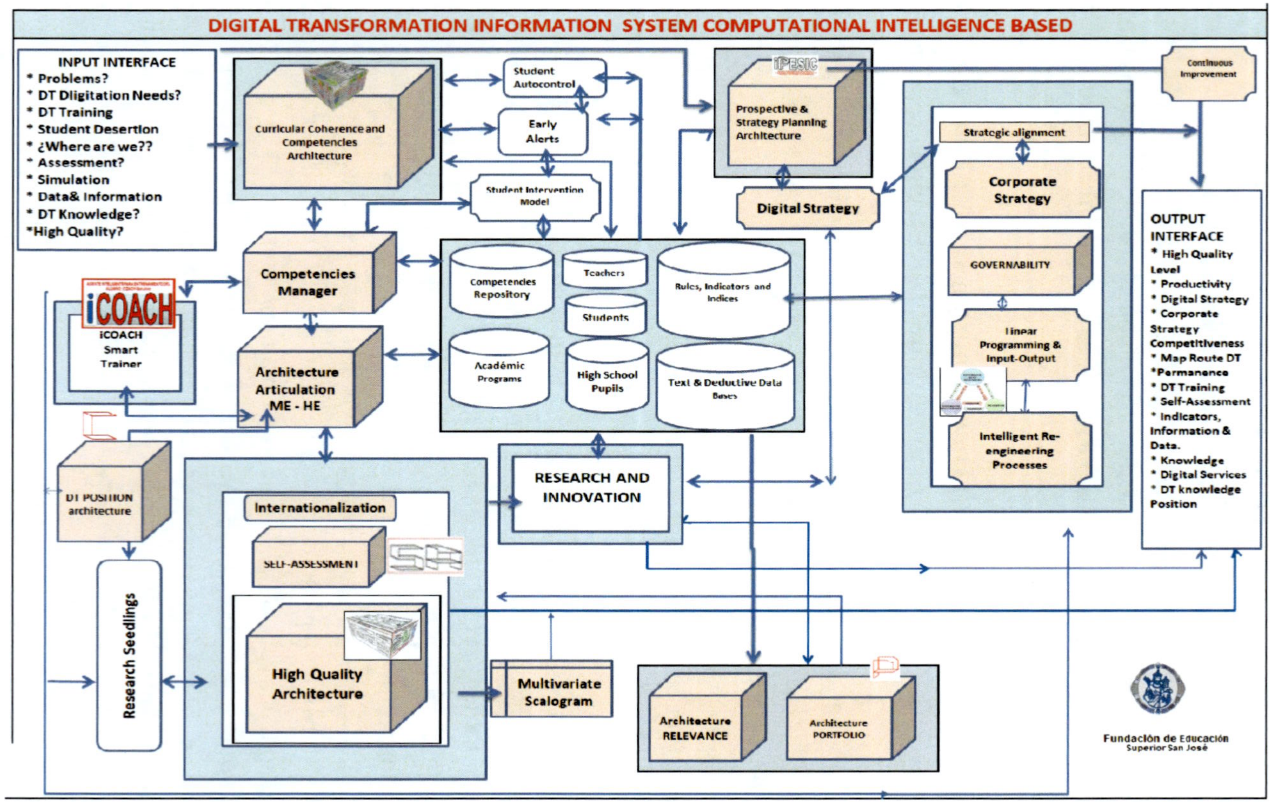

Figure 3.6. iMIS USANJOSE-ARKE-INTELL for DT.

The iMIS integrates hybrid multilevel system architectures, such as a DT space, to respond to the solution of HE problems tending to competitiveness, specifically pointing out how these new technologies can be included for their adaptation and evolution at the university in the modernity, making *governability*, *teaching* and *student productivity* compatible with *educational quality*.

Y (iMIS) = f {Desertion problem, coverage, performance, technological level, management level, user-plan}

The Y (iMIS) represents the output or results of the data, the corresponding *output interface*, and the results of the iMIS, such as: dropout values, responses and plans, decisions, information (e.g., *permanence*), required knowledge, solutions, inferred plans, policies and objects of *continuous improvement*, among others.

In the case of decisions, the degree of *structuralism* is involved, from the managerial to the operative aspects, which can be inferred from the problem or made explicit.

The level of databases and knowledge (deductive data bases) about the students at risk of defecting: is the possibility of identifying deductively, new knowledge from the base of data, such as characterization of new risk factors to manage the factual data recorded, and the ex-ante knowledge registered without making it explicit.

The *coverage*: implies the functions according to management and/or abstraction levels, academic program/semester, and the area covered by the iMIS, e. g.: registration and control, accountant, financial, human talent, stores, educational services, family parents, Ministry of National Education (MEN), and others implied.

The *performance*: refers to the managerial ability, inferred or explicit in terms of quality and academia.

Decisional reasoning in several domains: the organizational and the academic, information required for decision making according to the quality and level of the user and MEN standards; it also involves the determination of mathematical and non-mathematical instruments (e.g., econometrics, statistics, and organizational and computer culture of the institution).

3.6.2. Academic and Institutional Productivity Architecture

Specifically, this architecture indicates how the new technologies can be included both for the development and construction of the iMIS and for its evolution in the modern corporation, combining and strategically aligning governance and productivity *quality* towards the HE *competitiveness*. This architecture helps us to have a 360° for visualizing, diagnosing and quantifying, as we are approaching from now to the academic superiority. It is composed as seen in the following section.

3.6.3. Functional Architectures of USANJOSE-ARKE-INTELL

The iMIS integrates multi-hybrid system architectures as a space to meet the problem solving aimed at a competitive HE, specifically noting how they can include these new technologies.

There are 7 functional architectures:

- Academic productivity for curricular analysis, management and optimization.
- Student productivity with intelligent coaching.
- Building management technology: construction of the future: roadmap for student permanence.
- Institutional productivity process engineering for governability.
- Quality management for *high quality in education.*
- Self-Evaluation.
- DT Classment.

Table 3.2. PSEd productivity

A. Academic Productivity for Curricular Analysis, Management and Optimization

Architecture to Establish Curricular Coherence of Competences, Matrix Analysis & Competence Management.

The final purpose of this matrix is the optimization of a curriculum and competency architecture for analysis, quantification and optimization of study plans and monitoring student progress. The architecture matrices relate knowledge of specific subjects vs. target competences establishing their implications. Incidence and relationships are expressed in qualitative values (fuzzy). With units of academic credit metrics that enable competences measuring and production of each subject's contribution to the overall curriculum for which they are established. Optimization of curriculum is possible and also, particularly, of propaedeutic cycles.

A1. USANJOSE CURR-COHE: system to analyze and quantify the curricular coherence of the plans of study of the propaedeutic cycles in higher education. Certification of logical support registry - software Book 13-Volume 41-Part 205, National Directorate of Copyright, Bogotá.

A6. USANJOSE-ARQ-COHE-CURRICULAR: Intelligent Architecture of Curricular Knowledge and Competencies for the Optimization of a Study Plan. Registered certified software, National Directorate of Copyright: Book 13 - Volume 45- Part 200, Bogotá.

B. Student Productivity with Intelligent Coaching

ME-HE Articulation Architecture.

Intelligent Coaching on Basic Sciences.

Improves individual competencies through a smart coach supported by a knowledge base. The Student Self-Control Board and Early Alerts are individualized and supplemented with an Student Intervention Model supporting student productivity. It is complemented by an articulation architecture with the ME, iCOACH which is applied in *mathematics* and *physics* and verifies the individual progress of each student in relevant competences.

Intelligent Architecture of Curricular Knowledge and Competencies for optimizing a Curriculum in Higher Education: DNDA (National Directorate of Copyright) Book 13 - Volume 45 - Item 197, Bogotá, 2015; COLCIENCIAS, 2017. Management Guides: ISBN978-958-48-0412-9.

C. Building Management Technology

Tool for Future Analysis.

Building future: route map for student permanence. This allows the generation of the action plan (road map) on student permanence, the *strategic planning and prospective*, supported by a student's individualized intervention model.

Table 3.2. (Continued)

This construct contains all the knowledge of the plan and allows to navigate in time and space for management information towards future construction, enabling management scenarios and intelligent management strategies (i.e., with metadata and meta-knowledge). It is the analytics of the *technological management capacity*.
USANJOSE-PLAN-ESTRAT-PROSPECTIVA: Prospective and Strategic Planning for Student Permanence in Higher Education. DNDA Book 13 - Volume 45 - Item 199. Management Guides: ISBN 978-958-48-0412-9.

D. Institutional Productivity

Intelligent Process Engineering for Governability & Sustainability.
Management *governability in student permanence from engineering academic and organizational processes* is supported with linear programming models with multiple objectives (3 levels of students in good performance) & Input-Output Leontief model, implemented through field work allowing the analysis, quantification and optimization, including the iterative analysis of continuous improvement that is also supported with a set of performance indicators generated by the information system. Process management is the architectural foundation.

Intellectual capital. Quantifies institutional productivity.

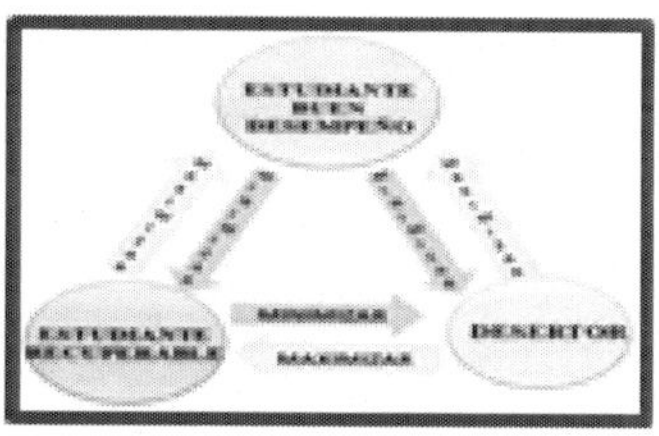

E. Higher Education Quality Management

Architecture for Governability of HQ.
Is structured by self-assessment analytical model with graphic display and is also based on a three-dimensional architecture: *dimension quality* of the student permanence *governability*, plus a one-dimensional entity (instruments). Among the tools are: Institutional Continuous Improvement and High Quality, towards the excellence. These architectures help us to have a 360° vision to visualize, diagnose and quantify, as we are approaching excellence now.

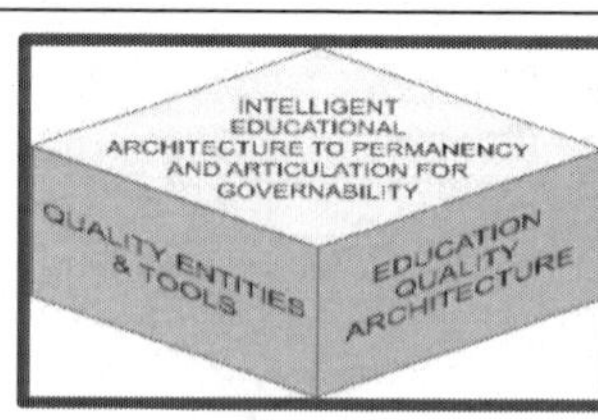

F. Self-Eval-Analytics

Intelligent Model of Institutional Self-Evaluation: is a complex architecture supported by the fuzzy logic of the AI for quantitative and qualitative *analytical self-assessment* of *research* and *innovation products* in front of different dimensions and multiple determinations of research knowledge management; it is a *mathematical-computational model* with AI and Fuzzy Dashboard Matrix Architecture that makes possible to obtain metrics of the university contributions. The developed Model is our commitment to the National Accreditation Council in DT. USANJOSE-SELF-EVAL-ANALYTICS - Intelligent Model of Institutional Self-Evaluation. Registered in the DNDA in Book 13 - Volume 65 - Item -52, Dec. 26, 2017.

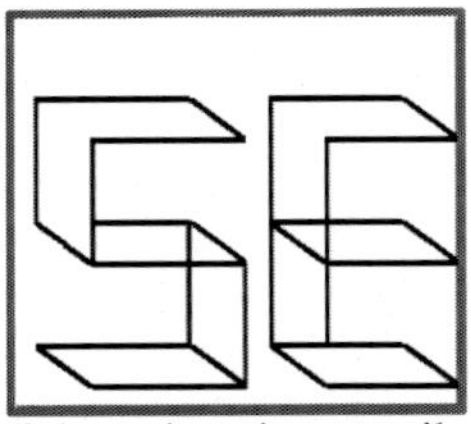

G. Architecture of Position Audit for Digital Transformation

Is a complex architecture supported by the *fuzzy logic* for the determination of the relative *level*, according to: evaluation of the digitization component, its *implications*, costs, degree of complexity, productivity relations, and income possibilities.

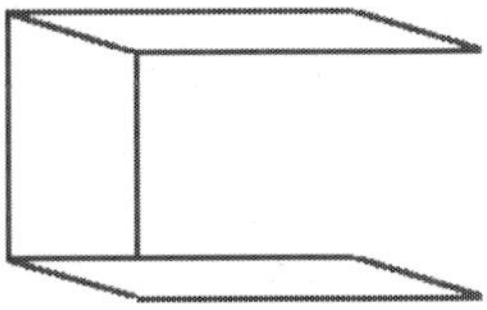

It allows classification of the maturity of *institutions* in relation to the DT, taking into account the possibility of evaluating and quantifying the multiple determinations of the DT as a system, the mentioned Ten Views of Our DT Model (Sec. 1.1), making abstractions from the DT Ecosystem and from *selected scenarios*. It was decided to consider 3 levels of maturity, taking into account also, that a characterization of companies was made to this effect, and reviewing numerous surveys and considerations of studies prepared. Our construct for DT classification, is carried out by *experts* who express their insights with fuzzy values.

Digital transformation degree of development. Is carried out on 3 dimensions, considering different components and establishing expectations about costs, revenues, etc.:

1- Dimension of culture: the focus of a university towards digitally driven innovation, and how it empowers digital talent and other employees (teachers, staff) with digital technology.

2- Organizational dimension: how the university is aligned to support the digital strategy, governance and execution.

3- Technological dimension: includes the infrastructure and the Web. The use and adoption of an emerging technology is standard.

H. Portfolio Analytics

Is a complex architecture supported by the *fuzzy logic* for quantitative and qualitative analysis of the product portfolio and its evaluation. It emerged with the idea of evidencing and quantifying the relevance with an architecture of our compendium or portfolio of final and intermediate products in front of the objectives of a call, sub-objectives, lines and sub-lines, using artificial intelligence with qualitative values.

Table 3.2. (Continued)

It makes possible to make quantitative and qualitative analysis of portfolios. The developments constitute the qualitative and quantitative starting point to explore extensions and potentialities of competencies in terms of skills, abilities, expertise, proficiencies, capabilities, etc., tending to develop curricula focused on the Knowledge Society in the 3rd Millennium, inherent to the *high quality of higher education*. 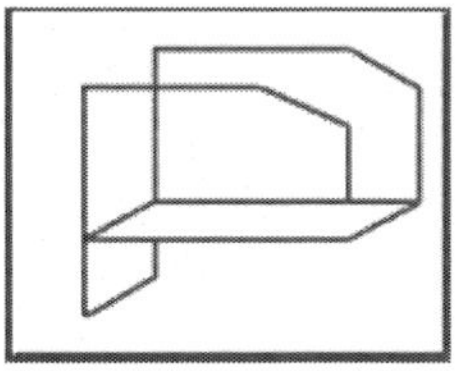 USANJOSE-PORTFOL-ANALYSIS: is a complex architecture supported by the *fuzzy logic* for quantitative and qualitative analysis of the product portfolio and its evaluation. DNDA Book 13 - Volume 58 - Part 17, Jan. 2017. Management Guides: ISBN 978-958-48-0411-2.
I. University Social Projection Analytics *Description is* an Intelligent model for University Social Projection and external and international relations. The social projection of the university is one of the fundamental pillars of our DT model. It requires the mapping of products, services, knowledge, consultancies and so on, society, industry, government and even the rest of the world. In synthesis, it refers to external relations, which obviously includes internationalization

3.7. DT Objectives

Generally, the DT objective is to extend the benefits of digitalization throughout the organization to achieve better competitiveness, academic and administrative productivity with the support of the academic and administrative strategies of transformation towards high institutional quality.

3.7.1. Specific Objectives

See below the Table 3.3 for specific objectives.

Table 3.3. Specific objectives

Objective	See tools references on Table 3.2
1. Digitize all the curricula in terms of its competences-objectives, measuring their impact on the universal knowledge of each subject with its academic credits and its micro-competencies.	A, G
2. Make the self-assessment of the institution's products as one of the instruments for migration and continuous improvement towards high educational quality that allows defining strategies for the reduction of operational costs.	F
3. Quantitatively analyze the product portfolio as impact of the research projects and of other activities on their social projection. In the same way, analyze the portfolio relevance in respect to the research lines and the results of the academic and administrative management. These analytical and deductive efforts should enrich all efforts towards high quality education.	H
4. Align the corporate strategies, the institutional educational project and the digital transformation strategies, taking into account the different normative and non-normative scenarios.	C, F
5. Develop an intelligent reengineering of processes towards their robotization.	E
6. Design and implement smart models focused on the students to improve their cognitive and academic productivity to reduce costs.	B, D, E
7. Design and implement intelligent models focused on the teachers to improve their productivity in teaching and in the development of knowledge verification methods.	B, D, G
8. Decrease the social gap in the human talent offer, for the world-class DT with social innovation.	B, F, G
9. Infer an institutional architecture with 360° vision that integrates all the efforts in productivity and quality, towards the competitiveness model objective, that consolidates the DT.	C, F

3.7.2. Intents and Purposes

Among the purposes it is worth mentioning:

1. Reach the new educational model based on competencies, micro-competences and high quality in the face of the new needs of the IV

Industrial Revolution, with new services and new capacities for generating knowledge and innovations.

2. Get the Mature Digital Marketing Model.
3. Repower the Model of Academic and Administrative Governability.
4. Redesign the Academic-Administrative and Enterprise Model of the Institution.

3.8. Tools for Future Analysis: DT Strategic Map

Strategic maps are methods and tools that provide a macro and simplified view of an organization's strategy and provide a language for communication to describe it. They constitute a graphic summary of the strategy of both the corporate and the strategy of the DT.

The **scenarios** we are using for DT development: **scenario planning**, also called **scenario thinking** or **scenario analysis**, is a strategic planning methodology that is used extensively in DT progress to make flexible long-term DT plans: the long journey (see Figure 3.1).

Environmental scanning: we are also using this, one of the essential components of the global environmental analysis as a part of architecting methodology: university environmental monitoring, environmental forecasting, and environmental assessment, complete the PSEd global environmental analysis. This global environment refers to the macro environment which comprises industries, markets, companies, users and competitors. Consequently, there exist corresponding analyses for our DT Ecosystem design (see Section 4.4). We understand environmental scanning as 'the study and interpretation of the education political, national economic, social and technological actions and trends (i.e., IV Industrial Revolution) which influence a business, industry and socio-economic or even a total market. The factors, which need to be considered for environmental scanning, are events, trends, issues and insights of the different actors and stakeholder's groups. Issues as the IV Industrial Revolution is not a simple paradigm change: it is an *era change*.

It is very important to take into account that the challenge for universities is to have numerical and cognitive metrics to measure their progress, especially in the long transit to DT. It is well known that you cannot control what cannot be measured: it is the logic of our ARCH-INTELL, which we present in Section 4.6.

We present our simplified ***strategic map*** scheme is a way to communicate the digital strategy and invite to the promotion of the digital culture in which we are engaged. In the next Figure 3.7 is presented our DT visualization through the strategic map:

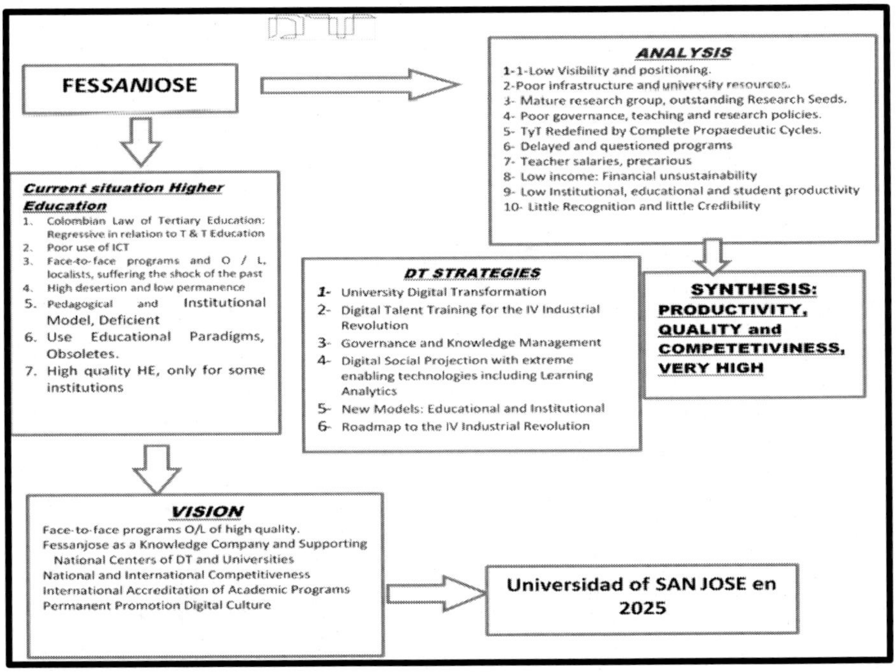

Figure 3.7. DT strategy map vision.

3.9. Computational Model

Table 3.4. DT computational model

MATHEMATICAL-COMPUTATIONAL MODEL OF DIGITAL TRANSFORMATION FOR POST-SECONDARY EDUCATION

(1) HIGH QUALITY ARCHITECTURE

HQ= ƒ (Self-evaluation, Membership, portfolio, resources, processes, internationalization, teaching productivity, institutional productivity $\mathcal{P}_i$, Student academic productivity $\mathcal{P}_e$, Academic productivity $\mathcal{P}_a$, Pedagogical model, Institutional capacity for generating knowledgeQ_k, Using Smart Tools)

(2) INSTITUTIONAL ARCHITECTURE

Economic branch production X_i: ƒ(Matrix $A^{input\ otput}$, $demand_{i,j}$, Académic Program_k, Student level, Intelectual Assets_i, $optimization\ policy$, $\text{Restrictions}_{i,j}^{k}$)

$\mathcal{P}_i$: ƒ(Competences-objetive, académic program.Tools (Edumatics builder, Test Design, Assessment))

(3) STUDENT productivity

$\mathcal{P}_e$: ƒ(iCoach, Alerts, Intervention, Architecture for Middle Education, Self-regulation)

Alerts: ƒ(Characterizations, level)

Student level: ƒ(Self-control, performance)

(4) TECHNOLOGICAL MANAGEMENT TFA

[Action Plan-Risk Analytics]]: ƒ(Internal-External Variables, Intelligent Strategies, Prospective, Scenarios)

Optimization of Opportunities
Threat mitigation

(5) Academic productivity

$\mathcal{P}_a$: FDM (, Academic program_k, Competencies, S_k, Knowledge (credits), Competences, S_k, Micro-competences μS_k)

Contribution of the Competences-Objective
Added value of each subject Knowledge C_k
Cost by credits N_j^k

Computationally, it is the expression that identifies in the iMIS USANJOSE-ARKE-INTELL, the dependent and independent variables and their interrelations, that act in all the space of the DT and facilitates the obtaining of the control system that interweave all the system.

3.10. DT Methodology: DT Development Process Based on Multiple Paradigms

DT development process based on multiple paradigms: models, architectures, systems, information, data and knowledge, regarding theoretical prototypes and real systems and our own experiences.

It must be understood as a very critical point of departure, that the DT development process is a long journey with multiple ports to arrive but that it does not have a final station as a definitive arrival (See Figure 3.1)

The multiple paradigms are related to the identification of functional and non-functional requirements within the academic and administrative systems to be integrated and published as executable constructs and services. The above requires:

- New profiles of users, services, processes, architectures, as to define new processes and the limits of information for those processes.
- DT updating and development of new tools.
- Interaction with different users and suppliers.

It is very important to:

- Understand PSEd objectives and define successful environments.
- Give importance to problems to be solved, then think on the technology.
- Be conscious of the solid requirements analysis of all institutions.

- Define the domain in which to circumscribe the problem scope of DT application in the organization (see the defined DT ecosystem).
- Face the small parts of the problem (i.e., subsystems).
- Understand all the application semantics that exist in the academic and institution domains.
- Understand the nature of all data, information, and knowledge that will be involved and exchanged.
- Identify possible circumstances and cases.
- Avoid unnecessary contradictions.
- Understand all the services available now in the domains: i.e., How many? Where are they? Goals? Objectives? Relationships?
- Understand all security problems, unauthorized access, privacy violations, weak protections and related.
- Automate the interactions between all digitization services.
- Be coherent and maintain synchronized the services and data/information/knowledge flows.

Preferably after, do:

- Select the technology.
- Deploy that technology.
- Test, verify, validate, legitimate and evaluate.

3.11. DT Methodology

The methodology refers to a systematic process and methods of reasoning, thinking in order to express and represent a problem and obtain an appropriate solution. It also refers to the way of planning and determining problems: it is the art of problematization. It is also the way to build a solution and at the same time verify that the solution is correct. Also, a methodology is defined as a collection of related processes, plans, methods, and tools used to define and solve problems.

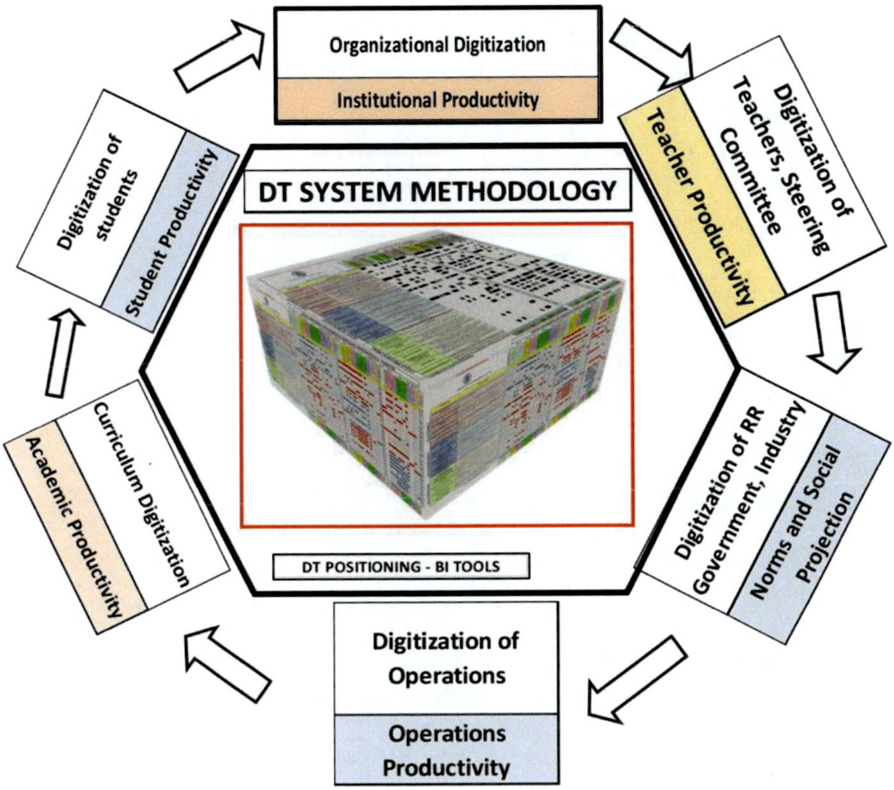

Figure 3.8. DT methodology.

Our methodology as mentioned before, covers the use of multiple paradigms ranging from information engineering, modern software engineering, data, information and software architectures, knowledge engineering with artificial and computational intelligence (ACI), and evidently WEB engineering, and others.

Planning DT processes, includes the determination of the solution system prior to fine-tuning the objectives to be developed, the detailed quantification of physical, human and technological resources, and the definition of security aspects over the Internet.

The study of the system for the diagnosis of the DT includes the requirements and processes engineering. The design involves the proven specifications of the system through evolutionary prototypes. The construction of the system is supported by tools for development of software of 4th and 5th Generation, such as tools like Qlik.

Table 3.5. Lineament

Lineament methodology	**Indicator**
System diagnostic study of the organization is updated. DT maturity is carried out. The organization chart is updated facing the new change of era and adaptations. The new digital culture is taken into account firstly. The Governability of the PSEd is redefined and updated. The **digital ecosystem** responds to the challenges of DT (See Section 4.4)	% Coverage
Computer Security is solid and includes analysis of the different risks due to the heavy changes and innovations that can occur with the DT development. Secure alignment of the Corporate Strategy (including the department ones) with the Digital Strategy (see Figure 3.7). The change management is carried out without losing the identity of the PSEd; preservation of the cultural identity of the PSEd. The definition of the DT Strategy, even with **limited budget** and digital talent: parametrization of resources.	% Advance
Study of users' habits (students, teachers, staff, parents and other clients) trends (relationship and satisfaction) should be digitized.	% Deepening
Infer IV Industrial Revolution Road Map, *vertically* and *horizontally.* Control: tracking metrics with Key Performance Indicators (KPI) & others. See Section 11.1.1 KPI: Key Performance Indicators.	Profile
Robotization of organizational and academic processes with knowledge management: identify and quantify key business processes and critical mission. Importance of each business process for the successful development of DT performance of the ICT applications that supported that business process. (See Section 4.5.3 360° Vision of Digital Architecture Functionalities.	Budget required
Previous definitions of the criteria for acceptance, validation, testing and verification of knowledge (legitimation).	% Advance
Analyze the competition, national and international, including the global market of online education.	% Advance

KE is used to represent educational/administrative knowledge and reasoning systems, organize the knowledge bases, their distribution and their communication, widening the bandwidth through hypermedia for fluid communication inside and outside the organization.

The short functionalities are described in Table 3.3.

Among the methodology guidelines in previous sections, the next Table 3.5 is a complement:

3.12. Some Other Extensions

3.12.1. DT Development Process Based on Models, Architectures, Systems, Information, Data and Knowledge

Methodology - stages are related to the *lifecycle*, which is the description of the distinct stages or phases through which each DT product passes during its system life. This includes phases such as requirements, definition, concept design, production, etc.

1. Understand PSEd EMM environment problems and objectives (see chapter 3) and define successful environments. See MIR Matrix.
2. Strategically:
   a. Customer experience (chapter 3) use the analytics: which is the use of quantitative methods to process knowledge like business intelligence (BI) & data to discover patterns, generation of knowledge, create new information and support decision making at all levels to provide such insights. Historically the term "analytics" is referred to processing structured data using solutions to provide such insights. However, for customer culture and experience, with the emergence of social computing and other Web-based technologies, a need arose to store and process unstructured data, like social media streams and comments, blog posts, and the like, with similar rigor.

   b. Use the *cloud* computing *with other extreme techs*, which is a type of Internet-based computing that provides shared computer processing resources and data to computers and other devices on demand, giving the variety of *users*. It is a model for enabling ubiquitous, on-demand access to a shared pool of configurable computing resources (e.g., computer networks, servers, storage, applications, and services), which can be rapidly provisioned and released with minimal management effort.
   c. Collaboration: working as a DT team to execute some process such as a design or engineering change approval. Asynchronous collaboration allows a team to work serially on something such as an approval, where the data being worked on is sent from one person to the next until the process is complete. Synchronous collaboration allows a team to work simultaneously on a process, seeing each other's data and comments, allowing the team to reach a consensus quickly.
3. The Ecosystem of the DT (see section 4.4).
4. The Structural vision:
5. The pillars of the *university* DT:
   a. These pillars are the transversal axes of very critical mission; they are: *research*, *entrepreneurship*, *digital talent*, *budget* and, especially, the *social projection of the institution.*
   b. The main components of the social extended projection in terms of *research*, *extensions service* and *consulting*.
   c. The *entrepreneurship* is widely accepted by its impact on the constitution of the DT model.
   d. The digital talent, which is very, scary need to be formed and trained urgently, given that right now is part of the great *digital gap*.
   e. The financial plan, because of the budget for DT, given that in the PSEd it's very restricted, and even more so, considering that it is an item that demands significant amounts in the short and medium term.

6. Sustainability: in the DT dual space for PSEd is included the governability. The DT sustainability model was presented Figure 3.4, where it is clearly shown the duality *academic space - administrative space*, whose characteristics are:
   a. The generation of new knowledge translates into publications of scientific-technical papers, properly indexed books, patents, vegetable varieties, software, or new animal breed, etc.
   b. The technical-scientific development and innovation is characterized by industrial prototypes.
7. The Strategic Map of the DT showing scenarios, actors and vision-mission, to be printed in flyers for DT diffusion as a part of the DT pedagogy (see section 4.8).
8. The architectural components for DT: the eight architectures developed and implemented for our DT model applying some intelligent constructs that we have developed and documented during the last 10 years at FESSANJOSE, leading the DT in PSEd (see section 4.13).
9. Digital architecture 360 ° of DT: Academy and Administration LOCUS (see section 4.5,4): this architecture is the *digital portfolio* and implies the organization of the subsystems to obtain better and/or new functionalities based on knowledge to obtain an intelligent behavior. The architecture-multilayer approach is a system of systems (SoS), which ensures compliance with government policies, norms and standards, in a highly complex social institution with digital assets; this approach describes the subsystems at a higher level, where a system is composed and with the protocols by which they communicate. It provides a 360° business vision map and a planning framework for commercial and technological changes.
10. The computational-mathematical perspective of DT identifies endogenous and exogenous variables, and interrelations (see section 4.9).
11. The synthesis: end-means matrix that summarizes in DT Where we are? Where can we go? (Section 3.4 - 3.6).

12. The MIR Matrix: DT Objectives, Impacts-Indicators and Results to be obtained with the DT. (see Section 3.9).
13. The dynamic model of the DT system based on *computational intelligence* (see 4.6) is representing the control system of all the components to achieve the completion of DT. It is an iMIS for PSEd that shows the dynamics of DT control integrating several multilevel, hybrid system architectures, as a space to respond to the solution of HE problems, tending to HEI competitiveness, specifically pointing out the way the modern technologies can be included for their adaptation and evolution in PSEd in post-modernity, making *governability*, *teaching* and *student productivity* compatible with educational high quality, the purpose of DT.

The interface of the Results of the iMIS are: *high quality metrics*, *indicators* and *values of management* for *desertion*, *answers* and *plans*. These results can be decisions, information (e.g., *permanence*), required knowledge, solutions, inferred plans, policies and objects of *continuous improvement* for the sustainability of the transformation and quality, among others.

The input interface is composed of: data, information and knowledge acquisition: where the attributes, parameters, values, problems (e.g., desertion) are registered and actively and dynamically obtain structure. The problem is expressed as the type of decision to be made, e.g., in the complexity of the *student dropout*, the need for information: i.e., students trained in mathematical competences, a prospecting of system variables, as rate of decrease of the gap *excellence-current state*.

3.12.2. Administrative Information Using Qlik

In this section, we present the Qlik Sense tool, with its central element, the associative engine that allows to discover knowledge that business intelligence and Educational Intelligence (EI) tools based on queries, simply do not capture. This democratization of information has enabled the power to deliver data in the DT processes, with a model of governability, almost in real time and with a concept of KE and management. The above, offers a number of possibilities that have leveraged the adequate exploitation of digitized information,

One case is for the users (students, lecturers, parents, government, industry): to maximize the service and optimize the management of the decision making by listening to the clients, understanding their needs and practicing what they preach, supporting mediation when the theories between the consultant and the company depart.

This process can only be optimized when all the actors align their efforts and can develop without being experts, in the diverse solutions of wide spectrum and multi-perspective orientation, diverse projects of search for answers through the data already being public and/or private, are able to give answers to the business in an appropriate way and with top quality graphics.

The different approaches, in a general way have limited to two dimensions the analysis of information, and that is when platforms like Qlik take the necessary relevance to efficiently guarantee the loading, transformation, development and deployment, duly supported in the understanding of the business, and finally the discovery of relationships hidden in the ocean of information, that will make the big difference. This is reflected in the transfer of skills through intuitive solutions oriented towards self-discovery.

What is sought is the transfer of business knowledge, taking advantage of collective (SWARM) intelligence, which is evident in fully agreed business rules, which facilitate the understanding of the environment in a common way.

In the first instance, its great differentiation focuses on getting connected to any database, linking the information of said information source through pointers, thus reducing the size of the database by up to 83%. making the reading and information processing much more efficient.

On the other hand, it generates an own file that works in RAM memory, which allows to read the request much faster, since it does not have to access the disk as such and therefore does not consume server resources. So, given the information is related, it allows to show (through its color code, -green, gray and white-) both the information that is related, as well as the information that is not, achieving a 360° architecture view of all information.

A typical example of this process is the customers being marketing to, during this year, and those who do not; a situation that requires a new query to fulfill its mission under a jeopardized cube search.

This associative model of information, in principle is not easy to understand, but it is highly intuitive for the mental process, bringing the technology even closer to the end user, so it is free to search, explore and organize according to what they require, not what technology in an isolated way allows them; in a habitual way, the process appears like this: ask, wait to answer..., because the resources of support for the generation of diverse consultations is limited.

On the other hand, rule-based security for granular control, data access and centralized administration allows generating the necessary governance for hundreds of users, whether in web or mobile environments, complying with all the requirements of governance, security and scalability in ICT.

Giving that, its community of users, allows the adequate exchange of developments and extensions built by the external software developers themselves, which allows an unlimited development based on the business, not the developer as such.

Within the plans one can see the interaction with a SDK (Software Development Kit), which will facilitate the development of institution solutions and leave aside the reports, queries and graphics, so demanding for the development teams.

It even has a Qlik web testing environment that, although very limited, allows to work with simple databases that require little space and processing at no cost. According to their route map of 2018, the developers has promised versions of the called QIX engine that will be able to connect with the DB of other applications, which makes it an excellent option for the open source by components, in the best open SDK style.

3.13. PSED Federated Data Architecture (FDA)

FDA is a form of system structure that accommodates multiple heterogeneous databases: academic, institutional, scientific, etc. FDA recognizes that information required to execute DT processes usually resides in multiple repositories, which often represent a completely heterogeneous mix of technologies and implementation approaches.

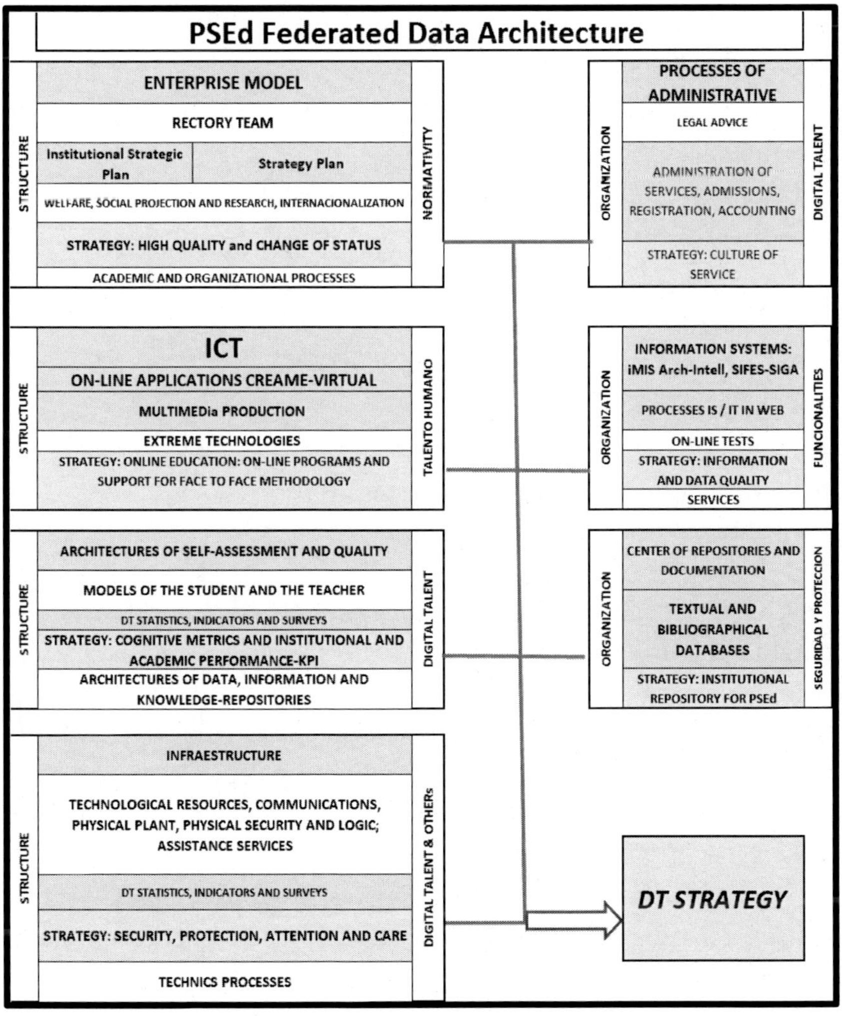

Figure 3.9. PSEd Federated Data Architecture (FDA).

Document Management including *data text bases* is set of the processes, procedures and tools focused on creating, maintaining, and managing documents. Comprehensive applications are available that provide authoring, scanning, structuring, interpretation, and printing of documents.

A business model is an abstract schematic representation (including the mathematical one) of an organization, to represent its structure. It also includes the value proposition either textually or graphically, and the central portfolio of products or services that the organization offers based on the actions necessary to achieve the strategic goals and objectives. It reflects the rationality as a system that creates, delivers and collects value in social, technological, cultural and economic contexts, etc. It is considered the basis of planning to build the business future.

Another way to see the institutional model can be seen with the canvas shown in Figure 3.9 that we built for our San José University.

3.14. Business Model

Our business model is the methodology for creating, delivering and capturing value of our research. See Figure 3.10.

Our value propositions are:

High quality education with *learning analytics* to compete in the IV Industrial Revolution with a *new educational model* based on custom-made international competencies.

A fundamental instrument is strategic planning and foresight. We have built our own tool that can be used centrally or by departments. Likewise, self-evaluation is used by government regulations. The strategy is to obtain HQ and change of status.

On the other hand, we use ICT in its two edges: *online education* (with support of face to face educational mode), and support of all the administrative users.

The component of HQ and self-evaluation is supported by our tools and quality architectures. In this case, the strategy is the production and use of cognitive and scalar metrics to monitor and control progress.

The ***Qlik* tool** has enabled us to use several legacy information systems. A system that we are building is a documentary repository digitized as institutional memory.

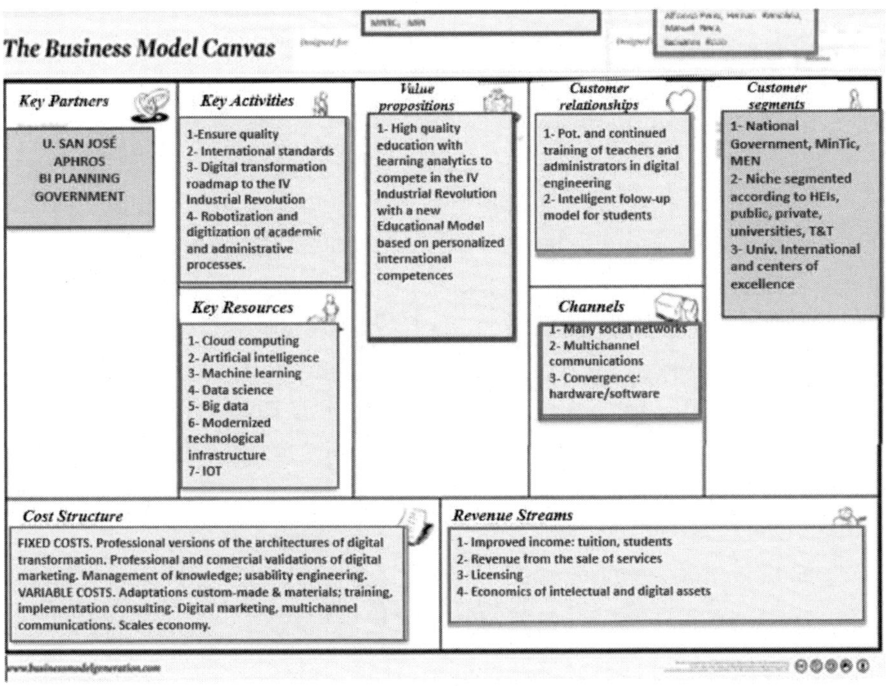

Figure 3.10. USANJOSE DT business model.

3.15. Final Remarks: From Computerization to DT, a Better World?

3.15.1. DT Started Several Years Ago and Is Consolidating

- Many evidences show that 30% of investment returns is feasible.

- The digital challenge is the new business culture based on digital skills.
- Current millennial users deserve special attention: it is necessary to listen to them, who will see everything and educate with mobile phones. The social and the economic are being redefined, in which the daily life will change, including promoting changes in the world order, economically and socially.
- DT focuses on business, prescriptive analytics, agile and modern infrastructure, staging capacity, improvement in the use with dynamic IOT platforms and cloud-native apps.
- Use of the Modern Analytics: How to generate greater value through the data, wherever possible? At any opportunity? It includes: data integration; unified data-types; and involve different sources.
- DT develops a readable hyper-plan (layout) of the *architecture*. Develops it in house internally or by outsourcing contract.
- It makes intense use of management and performance metrics and indicators; e.g., applies ROI, conversion rates, qualified talent, condition of marginality, organic rate of repurchase (without investment in marketing), and SLA. Permanently evaluates the virtualized experience of the user.
- DT is not reengineering neither retro engineering: it is digital mapping, extended horizontally and vertically.

Our Great Chapter Conclusion

The DT of our society changes the way we live, work, learn, communicate and collaborate. This disruptive change interacts with all the processes and systems that are important enablers of activities for the DT for years.

We contribute to the literature by presenting this new perspective for adaptable digital educational architectures for the university with our **Dual**

DT Model: curricular digitization and institutional digitization, i.e., knowledge and data.

We presented the DT with 10 different visions that allowed us to understand the DT through different worlds, with which we recognize the high complexity of the DT. The dynamic vision is represented with the iMIS for the search of the DT towards new high levels of scientific-technical maturity to achieve the desired levels in high competitiveness and quality. It is our work of many years knowing the HE sector, visiting numerous universities in our condition of Academic Par, as delegate of the MEN.

This has been enriched with our San José EIDOS Research Group, where we address very complex HE problems such as student desertion and permanence, quality, productivity, costs, propaedeutic cycles, and competitiveness at several levels, among many other problems.

As a research activity within our San José EIDOS Group, several constructs of both academic and institutional digitization have been produced, duly documented.

Our models and architectures (AI based) of Self-evaluation, Analytics of portfolio, High quality, DT methodology, and DT ranking, are other constructs that we offer to the academic community.

This chapter allows us to conclude that DT is a long-term and very complex process, which is the only way for an HE institution to survive successfully in its transit towards the IV Industrial Revolution.

Chapter 4

PORTFOLIO ANALYTICS

ABSTRACT

Architecture supported by AI Logic Fuzzy for quantitative and qualitative analysis of product portfolio, is a system that formalizes reasoning that contain fuzzy statements, commonly used in natural language, arguments that classical logic cannot address but that have wide uses in AI. It implies the fuzzification or granularity of truth, with ambiguities of sets whose limits are not well defined. It is a diffuse inference mechanism, based on a knowledge base.

Keywords: federated data architecture, portfolio.

4.1. INTRODUCTION

USANJOSE-PORTFOL-ANALYSIS: is a complex architecture supported by the logic of *fuzzy predicates* (See Chapter 1) for quantitative and qualitative analysis of a product portfolio and its evaluation. It seems obvious that fuzzy logic is a system that formalizes permitted reasoning containing statements with terms called fuzzy, of very common use in natural language, reasoning that classical logic cannot address but have broad mathematical uses in *artificial and computational intelligence*. It implies the "fuzzification" or "granularity" of truth, handling ambiguities of sets whose limits are not well defined. Fuzzification is a diffuse inference mechanism, which starts from a knowledge base (rules). A fuzzy predicate is one that presents "border cases" as illustrated in Figure 4.1.

4.2. Origins of the Portfolio Construct

Our construct emerged. as a result of the Convocation 745 of COLCIENCIAS, resuming research work and previous experiences of Convocation 755 MEN – FESSANJOSE Agreement (2011-2013), with the idea of highlighting and quantifying the relevance of our proposal, with an architecture of our compendium or portfolio of *research*, final and intermediate products, in front of the objectives, sub-objectives, lines and sub-lines of the convocation, using FL of AI with qualitative values and non-numerical mathematics. The construct makes possible to do quantitative and qualitative analysis of portfolios. The developments constitute the qualitative and quantitative starting point to explore extensions and potentialities of the objectives of quantitative goals, objectives and sub-objectives, impacts, incidents, etc., of the portfolio, tending to develop projects focused on the knowledge society in the 3rd Millennium, inherent to the *high quality of a research and innovation portfolio*. The construct is implemented in FESSANJOSE-Bogotá with all the documentation, under the IEEE standards, and also includes a *handbook* User Manual for the System Operation of the USANJOSE-PORTFOL-ANALYSIS.

This *analytical portfolio and relevance matrix* allows to visualize with clarity the relationships, incidence or impact that there are between the *goals, purposes, objectives and sub-objectives* of a *portfolio of services, products, goods and assets* offered, averaged from the evaluation that can be done by 4 different experts in qualitative (and linguistic) terms belonging to the *fuzzy* logic of AI already mentioned.

4.3. DESIGN SPECIFICATIONS: THE MATRIX ARCHITECTURE

In general, we use our FDM matrix (see chapter 2) that is MxN where:

- The M rows represent the levels of the purposes and objectives to be obtained from the *portfolio*.
- The N columns represent the unitary and functional areas of *the products, goals* and related.
- Each coefficient (cell) $a_{i,j} = \{0, 1\}$, for the case of a binary matrix is 1 if there is direct involvement between the area of expertise and each subject of the curriculum of the cycle. Or is 0 otherwise.

On the other hand, each coefficient (cell), which represents the cohesion and coherence of each objective with the production, can contain other wider qualitative values (fuzzy logic) so that the degree of possibilities is increased.

$a_{i,\ j}$ = *{null, low, medium, high} (white, green, red, blue of the* Table 1.2*)*

4.4. FUNCTIONAL REQUIREMENTS OF USANJOSE-PORTFOLIO-ANALYSIS

1. Study the relevance and interaction of the set of products, services and results of a project (*outputs*) against the *objectives, purposes, goals, impacts* and *purposes* (*tickets*).
2. The interrelation is qualified with qualitative values such as: high, good, average, regular and null.

3. The above is mediated by *experts* who qualify the incidence of the exits in front of the entrances. The intelligent tool enables the automated intervention of several experts and their registration.
4. The FDM Matrix of Pertinence is completed in several directions: *top down, bottom up*, or standing at an interior point, providing a 360 ° view.
5. The analytical space that is provided makes possible the optimization, via the cognitive metrics obtained through the defuzzification process.

Defuzzification as mentioned above consists of obtaining a numerical value from a fuzzy value or fuzzy set. That is, about choosing the number that better or more usefully represents a certain distribution of possibility or also the degree of membership to a set.

There is no unique *defuzzification* methodology to make this choice. The three methods currently used are: maximum method, maximum average method and method of the center of the area or weighted average. In this last case, we weigh the costs of each *product*, n_j.

4.5. FUNCTIONAL SPECIFICATIONS: CALCULATION OF DEFUZZIFICATION

The following relationships are obtained:

$\sum_j a_{i,j} n_j = b_i$	Total of the objective i to be reached, weighted by the costs associated with the product j.
$\sum_i a_{i,j} n_j = d_j$	Total of the contribution of the product d_j, weighted by the costs associated of such product n_j.
$\tilde{a}_{i,j} = \frac{d_j}{b_i}$	Technical coefficient of the objective d_j per unit of production b_j.

Figure 4.1. Defuzzification relationships.

4.6. ARCHITECTURAL SPECIFICATIONS

The tool must handle multiple layers.

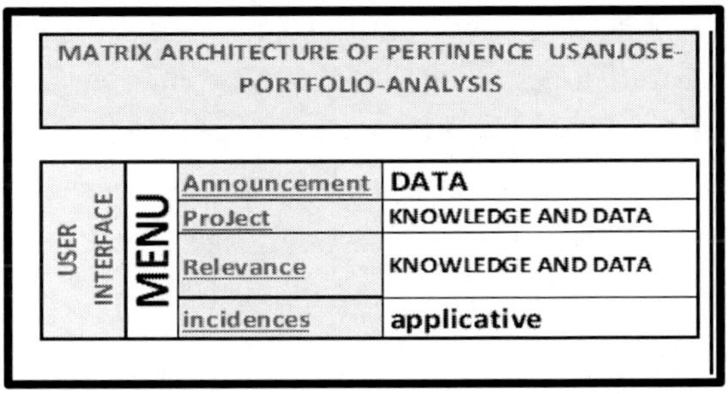

Figure 4.2. Portfolio menu.

4.7. NON-FUNCTIONAL SPECIFICATIONS

The tool must comply with the commonly accepted IEEE standards for end-user computing:

- The usability acceptance criteria.
- Allow audit trails.
- Enable quality assurance.

4.8. DESCRIPTION OF THE USANJOSE-PORTFOL-ANALYSIS PROGRAM

The program contemplates the following processes:

1. It assumes and counts on the knowledge on the part of the user about the basic operation of the Windows operating system and Microsoft Excel, in general and in common tasks from the start-up, including

selection of options; the details of the implementation can be found in FESSANJOSE-Bogotá with all the documentation, under the standards of the IEEE, and the *handbook* of User's Manual for the Operation of USANJOSE-PORTFOL-ANALYSIS System[15] (windows, sheets, cells, digitation, etc.)

2. Presentation of a set of spreadsheets by tabs in a book so that the user can access any without a pre-established order, but shown in the most natural order to carry out the activities which they represent; one of the sheets is "Menu", an introduction and guide for the use of the application using hyperlinks, and to which the user can return at any time.
3. Modeling of a procedure to process a call and a project that is presented to that call: the program receives the entries of the constituent elements of the call, projects through spaces (pages) dedicated exclusively for this and distributes the information towards the sections (sheets) for processing and presentation.
4. Deploy matrix through the ordered cells, the model of the call, the project, and how they interact from their main constituent sub-elements: objectives and products.
5. Involve the participation of several peer external expert evaluators (PEEE), distributed through rows destined for each of them, who are responsible to give values of levels to each of the relationships mentioned in the previous point; and present to the evaluator the list of options to choose from differentiated by colors.
6. Fuzzification/defuzzification of the information to interact with the PEEE in natural language, process the data and present it again in natural language to the user with the same color code.

---

[15] Compendium of Complex Architecture supported by the Fuzzy Logic for quantitative and qualitative analysis of Portfolios of Products and Results with Management Guides for evaluation of relevance (Spanish) ISBN 978-958-48-0411-2.

SANJOSÉ

FESSANJOSÉ - EIDOS Group *USANJOSÉ-PORTFOLAN V2* San José EIDOS Group's Portfolio

Project: Portfolio Analysis

PRODUCTS AND SERVICES MORPHOLOGY

General characteristic: Portafolio de productos y servicios de investigación al Grupo San José EIDOS

Main reference: EIDOS Group's production document, 2018

NRL1	TTL1	DSC1	RFR1	NRL2	TTL2	DSC2	NRL3	TTL3	DSC3	PS
2	Books			19	ARQUITECTURAS DE DATOS, INFORMACIÓN, CONOCIMIENTO, TIC Y DE SISTEMAS DE INFORMACIÓN GERENCIAL CourseWare	20 B6. Ing. Alfonso Pérez Gama; Universidad Distrital Francisco José de Caldas – II Edición 2009; FESSANJOSE y ACCIO Asociación Colombiana de Informática 2009. Libro Electrónico ISBN 978-958-44-5164-7 – Bogotá. CONVENIO FODESEP 2009.				1,00
2	Books			20	Tecno-Dialecto Especializado de Cognimática, Planeación Estratégica y Auditoria de Sistemas	21 B7. – ALFONSO PEREZ GAMA; ACCIO FESSANJOSE Bogotá. Mayo 2008. ISBN 978-958-44-2172-2 CONVENIO FODESEP 2008.				1,00
2	Books			21	COSTOS APLICADOS POR PROCEDIMIENTOS	22 B8. HERNAN REMOLINA CARO; REGISTRADO en la DIRECCION NACIONAL DE DERECHOS DE AUTOR, LIBRO 10-TOMO 330- PARTIDA 360. Fecha 26-May-2012, ISBN 978-958-46-1535-0; Disponible en WWW. AMAZON.COM				1,00
2	Books			22	Compendio de Arquitectura Inteligente de Conocimientos Curriculares y Competencias para optimización curricular con Guias Académicas para optimización.	23 B9. ISBN: 978-958-48-0409-9 BOGOTA 2016				1,00

Figure 4.3. Partial view of portfolio matrix 1 (products table).

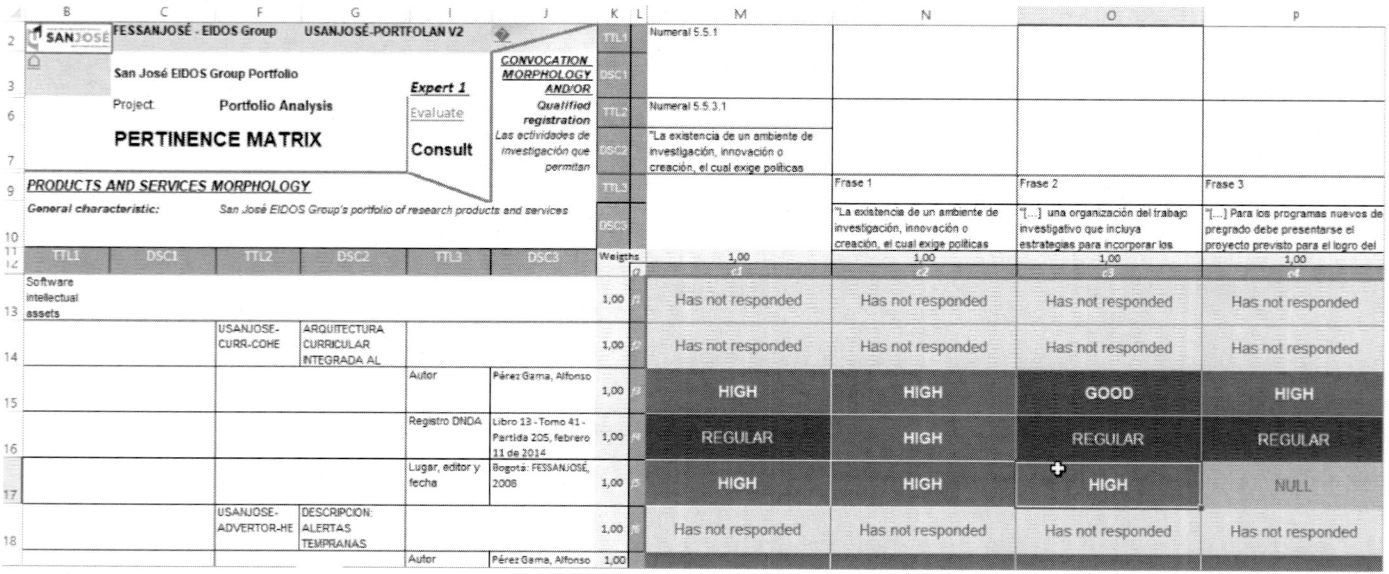

Figure 4.4. Partial view of portfolio matrix 2 (expert 1 matrix).

7. Processing the information going between the steps of conversion from natural language, that is, in its numerical form, and conducted from the evaluation sheet by the PEEE, to the presentation sheet, in the form of average/rounding calculations per cell, and matrix calculations of total weighting by rows and columns.

Here are two partial views of the FDM: *pertinence* view and the *incidence* view (Figure 4.3 and Figure 4.4).

CONCLUSION

The analytics of the university product portfolio, is supported by the AI (using an FDM), in a tool that allows us to permanently evaluate the progress of our DT, what nourishes us (feedback) and guides us with the anticipatory information of the future (feedforward).

Chapter 5

SELF-EVALUATION AND HIGH QUALITY EDUCATION ANALYTICS

ABSTRACT

This is a Mathematical-Computational Model with AI; a complex architecture with FL for analytical, quantitative and qualitative *self-evaluation* of *research* and *innovation products* in face of multiple determinations of research knowledge management. A Fuzzy Dashboard Matrix Architecture like this, makes possible to obtain metrics of university contributions. The developed model is our commitment to the National Accreditation Council in DT.

Keywords: self-evaluation, fuzzy dashboard matrix

5.1. SAN JOSÉ SELF-EVALUATION ANALYTICS

The self-evaluation precedes the high quality in HE. We present a *mathematical-computational model* supported by AI: USANJOSE-SELF-EVAL-ANALYTICS. It is a complex architecture supported by *fuzzy logic*, for analytical, quantitative and qualitative *self-evaluation* of *research* and *innovation products* in front of different dimensions and multiple management determinations of research knowledge. The basis of the *architecture* is a Fuzzy Dashboard Matrix (FDM) that allows obtaining metrics of the products and results contributions of the university work in *research*, innovation (RD&I), *teaching*, *social projection* and *extension*. Our proposed and developed model, USANJOSE-SELF-EVAL-ANALYTICS is our bet for the National Accreditation Council, NCA, within the line of *digital transformation*.

5.2. OVERVIEW

The institutional evaluation of the universities constitutes a tradition of new experience in Colombia since the General Law 30 of Education was promulgated in 1992 consecrating the *university autonomy*, although this had already been carried out decades ago in some countries of the Ibero-American region.

The regulation and formulation of complementary decrees needed for the extension and consolidation of the university autonomy was complicated by the multiplicity of HEI and, in another order, by those factors activated by social and economic changes (a war of more than ½ century), that require new solutions and responses from institutions that must interpret these evolutions.

Evaluation and accreditation constitute a necessary instrument of policy, planning and university management, towards the growing need to guarantee high quality of the HEI. In fact, the institutional evaluation is a complicated and highly complex process to establish the value of a system, which implies it must be free of subjectivism, worse situation when there is high content of objective evidence and interpretation of aggregates of elements that interact inter- and intra-system with purposes, goals, and concrete objectives, with the consequent great systemic implications.

In itself, to evaluate is not inflexibly the final phase of an interactive process system in the academic and administrative fields, but is a road map pointing away to appreciate the execution, and conclude how random and very complicated is the road to excellence.

5.3. Introduction to Institutional Self-Evaluation

Given the essence of their mission, HEI must always seek excellence, that is, carry out analytical studies and managements that allow continuous improvement. Analogously to living organisms, educational organizations develop by reacting to stimuli and problems to the internal and external environment, making the appropriate, effective and timely transformations. In other words, the organization must self-control to achieve the mission, vision and objectives.

A self-regulated educational institution must necessarily continually and systematically review the objectives of its tasks, establish an efficient information system based on facts and opinions about the different elements of the organization, act on the basis of scientific studies, and connect the results with the planning, the allocation of resources, and the improvement of the organizational and institutional processes.

Self-evaluation is mainly a mean or systemic instrument that the institution uses in the pursuit of excellence, closing the gap between its limits and the current level of quality. Self-assessment is also a process of study of an institution or one of its parts such as a faculty, a department, a service unit, or a program, which is organized and conducted by its own members, in light of the aims of the institution "and some accepted set of performance standards". The study has descriptive elements, but, mainly, it is analytical and reflective, and assumes a commitment to change.

5.4. The State Self-Evaluation: National Council of Accreditation, NCA

The NCA Model is very complex. These are surveys that must be done by institutions of higher education periodically. Its manual analysis introduces more difficulty.

It has:

- ***Factors:***

Between 8 and 10, which are groups that affect the quality.

- ***Characteristics:***

Between 40 and 42, are measurable properties that constitute the characteristic component of the model, the mode of expression of an institution or the quality of a program. These are general and specific predicates of whether an institution is an academic or organizational process or a service. For example, in the Student factor, the characteristics are:

1. Selection and admission mechanisms;
2. Admitted students and institutional capacity;
3. Participation in the integral formation;
4. Student and academic regulations.

- ***Aspects:***

Between 181-243, are relative values associated with empirical variables. For item 1, mentioned above, the aspects to be investigated are:

24. Special requirements for admission to programs that warrant it;
25. Admission policies and exceptional mechanisms that are made public;
26. Mechanism profiles supported by exception;
27. Evaluation systems and mechanisms, and admission and pre-selection processes, including the implementation of the results of this evaluation;
28. Admitted students of the last four cohorts in each admission process, through the application of general rules, or through the application of exception mechanisms.

The quantification mechanism is carried out through online surveys and guided interviews administered to all the actors involved (students, administrators, teachers), and through the analysis of data and information systems, at the *university* or the government institution.

For this business information architecture, we developed a multivariate matrix called Scalegram (see Figure 5.1) with fuzzy values for self-assessment (degree of integrity): the size of the self-assessment matrix for a year is 230 rows (aspects) by 45 columns (academic and administrative processes). We developed the matrix from one qualitative model to another with metrics, as we applied the same methodology of *curricular coherence architecture*, a previous work. The process of AI known as defuzzification that we have used many times, can quantify the self-assessment. In the near future, we can extend it to intertemporal analysis.

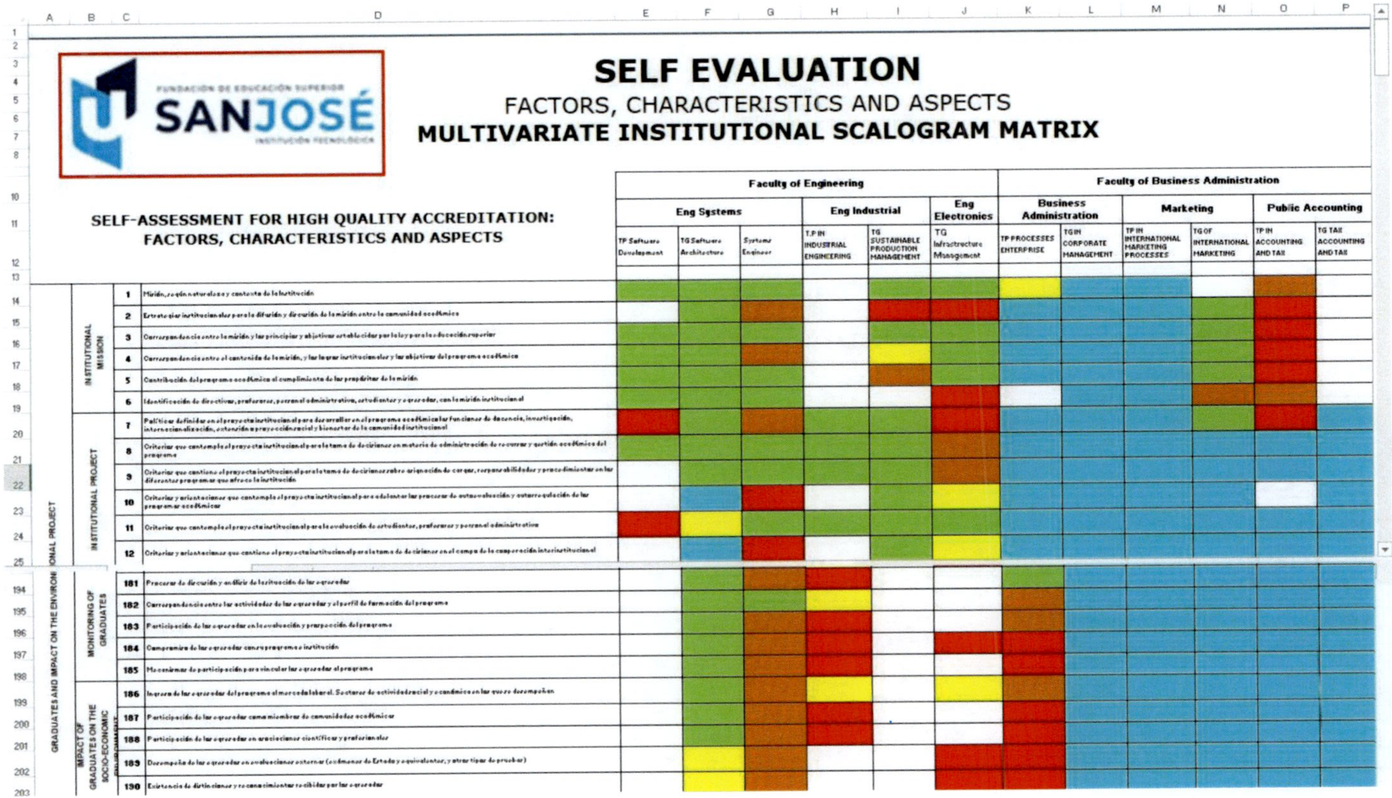

Figure 5.1. (Continued).

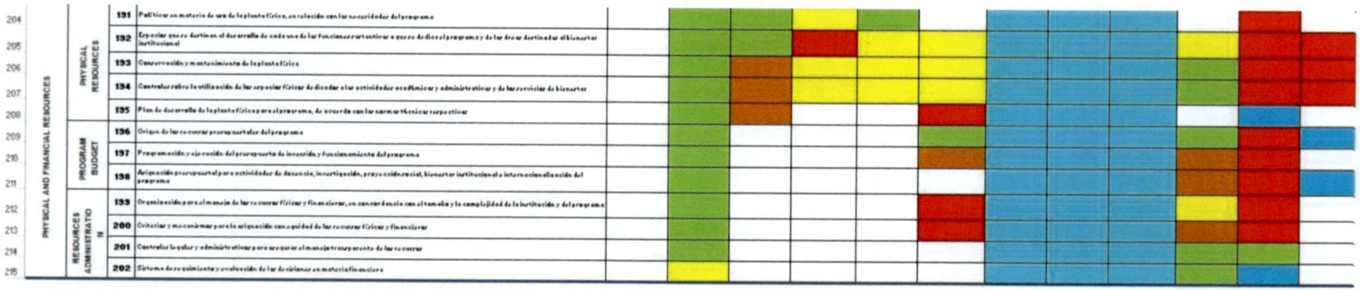

Figure 5.1. Scale of self-assessment NCA: Partial view.

5.4.1. Goals of the NCA Self-Assessment

The objectives of the institutional self-evaluation are to:

- Establish quality as *imperative* in HE
- Generate continuous improvement of processes
- Create critical thinking for change: knowledge of society
- Project new and improved academic and administrative processes
- Evaluate and maintain the quality of services to the internal and external users (students, teachers, parents)
- Make continuous progress in developing plans with national and international visibility
- Unify the processes and procedures effectively and efficiently, looking for organizational productivity
- Determine self-assessment as organizational processes (academic and administrative) to integrate it into the large Process Matrix for optimization purposes using our knowledge architecture
- Analyze and evaluate information in a well-structured 360° vision, which the architectural approach allows us to do, making possible to navigate all the administrative and academic information of the *university* (ascending and descending paths)
- To a large extent by the applied method, facilitate the strategic alignment of the different units, with the final strategy: *high quality*
- Through the information congregation, reduce the complexity of the analysis and the design of coherent plans for institutional improvement
- In the self-assessment administered as an organizational process, facilitate organizational metrics using linear programming and Leontief's input-output models (which extend their value)
- Integrate self-evaluation as a process variable. This allows to quantify the incidence of self-evaluation in terms of quality, academic sustainability, student desertion, and many others.

5.4.2. Self-Evaluation as a Process

Self-evaluation is also an essentially competent and rigorous process, although it responds to defined policies, and is related to the technological capacity of the organization to diagnose and solve problems, and create strategies to introduce, administer and sustain cultural, organizational and academic changes.

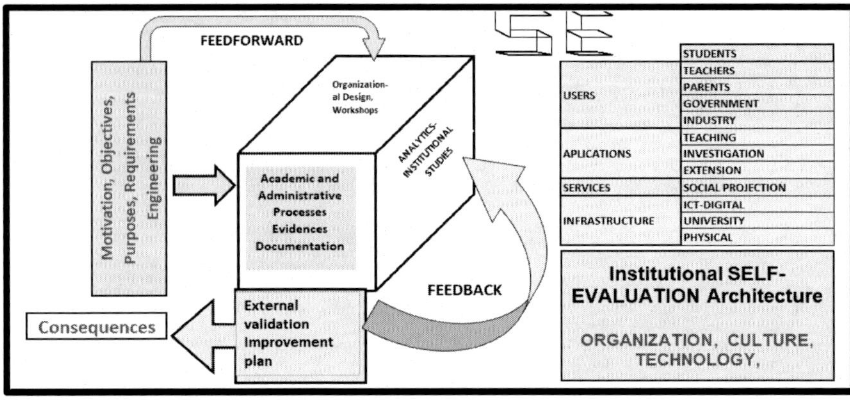

Figure 5.2. Architecture of the institutional self-assessment: Components.

It is a *permanent process*: the exercise of institutional self-evaluation is a systematic and rigorous continuous process that requires a system for collecting and organizing data and information throughout the period. At the end of the academic period, an analysis and evaluation of the recovered information is carried out to identify both the internal areas of greater strengths that must be maintained, and the weaknesses that need to be minimized and improved in order to obtain the expected results in the *high quality education*.

The institution has reached its level of development that already involves this logic of the strategic planning and the continuous improvement, including evaluating its internal, external, academic and organizational processes, which own of its mission that are consistent.

5.4.3. Appropriation of Knowledge and Best Practices

With appropriation of knowledge, there is evidence of the HEI level of development, and best practices have been known in the university environment, widely disseminated, and recognized, by the educational community, the government (MEN), parents, and the industry. Characteristics:

- Relevance.
- From the research lines and the institutional challenges, there is articulation of the institutional educational project (IEP) and the needs of the users and the external demands.
- Participatory: the process of institutional self-evaluation is fundamental to have the participation of students, parents, and teaching and administrative teams.
- Coherent: coherence responds to the consistency between affirmations and denials throughout the institutional self-assessment, and also to the consistency between the different components, and must respond to explicit and commonly accepted evaluation criteria, i.e., the tools, the relevant items, and the Self-assessment methodology must be known and internalized by the educational community.
- Validity: the results of the institutional self-assessment need to be recognized as truthful by the different levels of the educational community. Validity implies that a *peer* evaluation with the same methodology, information and expertise must reach the same conclusions and recommendations.

5.4.4. The Purpose of the Self-Evaluation Process

Self-evaluation is underlying in other purposes that can be established; it is the improvement of the *mission* and the work of the institution. The self-assessment can also be carried out in order to respond to the requirements imposed by the *high quality accreditation*.

5.5. Requirements and Conditions

The functional and non-functional requirements of self-assessment are described below:

- The institutional evaluation should tend to create favorable conditions (organizational climate) so that the participants, including the evaluators, could improve their knowledge about the institution.
- The purpose of the institutional evaluation must serve to unravel, change and continuously improve, so that the self-evaluation fulfills its objective of tending towards an improvement of the *high* quality as a fundamental premise. It is not about normalizing, prescribing, or let it alone as a "corrective or disciplinary activity". To do it like this, it must be done in a permanent and participatory manner.
- Support of a system that is retroactively reported, consistent, coherent and rigorous through a management information system, whose self-assessment report is a quantitative and qualitative presentation (i.e., with metrics) that displays the academic, administrative, students and professors' activities, extension, social projection, the organization and dynamics of the institution, as well as its goals, objectives, vision, values, policies, strategies, and educational plans, with their evidences and findings.
- Although the high participation of all the university levels is desirable, it is desirable that a group of experts with a required profile and competence make the objective evaluation (i.e., neutral). The former imposes the democracy of the arithmetic through the tool and avoid the biases of interest groups.
- The above also requires assuming with ethics, honesty and frankness, that there are values that are appreciated.
- The self-assessment report must adopt a contextual and historical perspective (diachronic and synchronic) of the institution referring to its IEP. The report must be very professional: rigorous and

concise, and orderly both in its purposeful design, and in its methodological aspect.

- Coverage of all the functions that are specific to the mission/vision. The *institution* has its systemic implications: HEI are complex organizations with multiple levels that interact with each other and with the environment, with social responsibility and their own identity.
- It is an ethical and responsible evaluation. The external evaluation with an expert endowed with appropriate competences must observe independence and respect the confidentiality and sensitivity of all institutional data. However, it is to be insisted about it as a process of exchange, and of open doors, which should allow a gradual (and eventually asynchronous, multistage, non-simultaneous) development of the implementation of the institutional evaluation in the different contours and levels of the HEI.
- It must also methodologically allow an adequate relationship between the specificity of details, the particularity (academic component) and globality (the whole of the university), seeking a balanced commitment to avoid the Manichean segmentation of the organizational and academic processes.
- Highly desirable flexibility, such that the different components of the institutional evaluation can be redefined and recombined over time, depending on the experiences and the changes in internal and external needs.

5.5.1. Structural Requirements

Next, the structure that abstracts the components of the architecture is presented.

5.5.1.1. Computational Mathematical Model

The left part represents the external or autonomous variables and the right part the results, products or *endogenous* variables:

Self-Eval Computational Model
Y = Ƒ (Model, Design, Team, Human Talent, Workshops, Reflective Analytical Study, Academic and Administrative Processes, R + D + i Production, Patents, Social Projection, Publications)
Y: *Results*
Reports recommendations, documented evidences
External validation
Improvement plan

Figure 5.3. Computational model of self-evaluation.

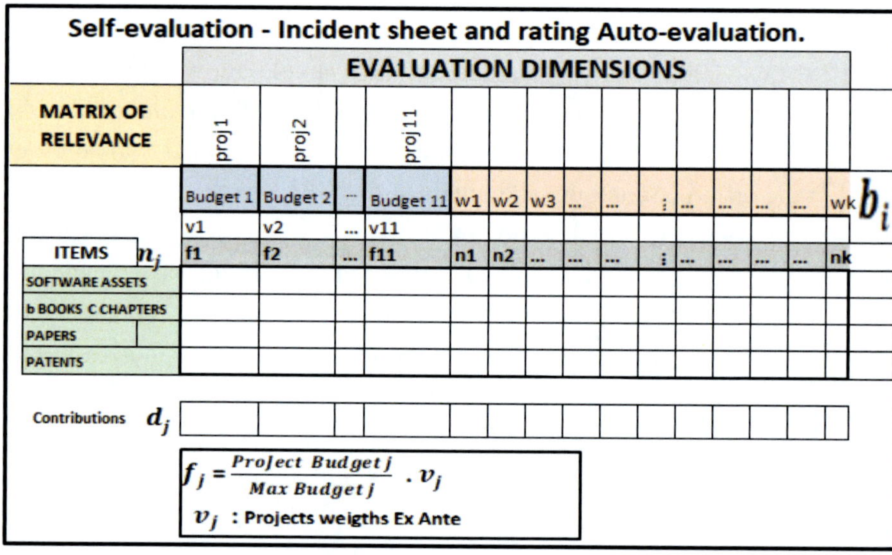

Figure 5.4. Self-assessment mathematical model.

5.5.1.2. Schematic Model of Self-Evaluation: Computational Morphology

The morphology is translated into the matrix arrangement between university products: knowledge products, publications (books, book chapters, papers), software, patents, prototypes.

5.5.1.3. Elements of the Model

bi: is the i[a] research production expressed of each item of the Group's Portfolio To obtain the fuzzy values $a_{i,j}$, in the opinion of the Experts,
which forms the Knowledge Base - Fuzzy **A**

$$\sum a_{i,j} \cdot n_j = b_i$$

j = 1, 2, .. 11, (projects), 12, 13 …k Assessment dimensions
i = 1,2,3… ITEMS TOTAL

where: **n**j are the factors of metrics, **d**j contributions of j[a] item
For each one of the weights $n_j = w_j \cdot r_j$
r_j **: normalization adjustment**

$$\sum a_{i,j} \cdot n_j = d_j$$

Represents the value of the contribution of each evaluation dimension j.
i = 1,2,3… ITEMS TOTAL
j = 1, 2, .. 11, (projects), 12, 13 …k Assessment dimensions

Figure 5.5. Elements of the self-evaluation model.

5.6. Quality of HE and Self-Evaluation

5.6.1. Quality as Perfection or Consistency

In *software* it is well known that a quality program has zero code defects (conformation to requirements specifications and design specifications, which are predefined and measurable). And other principle is: doing things right from beginning to end.

Excellence or perfection is to ensure that everything is correct, without errors. Yet more, this approach requires that such perfection be delivered in a consistent manner. The concept of excellence involved in this approach also implies a philosophy of *ex ante* prevention (assurance) instead of inspection.

5.6.2. Quality as a Fit for the Achievement of a Purpose or Specified Functionalities

If something does the work for which it was designed, then it is said to be of quality, and if it meets standards, it is close to *high quality*. The user's perspective identifies quality to the extent that a product conforms to the user's specifications.

Self-regulation/self-control is not about specifying standards or specifications against which to measure performance. It is about making sure that there are mechanisms, procedures and processes that allow *ex ante* to ensure that the desired quality, however it is defined and measured, is effectively delivered.

5.6.3. Quality as a Process of Transformation

The transformative conception of quality is based on the notion of qualitative change. The transformation does not refer only to physical changes, but also implies cognitive transcendence. This idea of quality as a transformation questions the relevance of the product-focused quality approach and its aptitude for a certain purpose, while several problems arise, as we saw, when trying to adapt to the HEI the concepts of quality based on the product.

In contrast to many tertiary services where the provider offers the services already elaborated for the consumer (i.e., student), this transformation process is unique and negotiated on a case-by-case basis.

The same reasoning applies to research. The provider not only produces new knowledge but is involved in the transformation of a body of knowledge for a special purpose.

This leads to two concepts of transformational quality in education:

- The enrichment of the interested party, and
- The user's reinforcement.

A quality education is what changes cultivated skills (competencies) in the subject and therefore presumably enriches it. This notion of *plus* gives a transcendental meaning to this enrichment:

1. The "added value" is a measure (quality metrics), as long as the educational experience enriches the knowledge, skills, abilities and skills of the students. Thus, a high quality institution would be one that greatly enriches its students. Knowing how much value has been added depends on the methodology, and what is defined as value. When the "added value" is measured, i.e., in cognitive terms at the input and output of the system process, a "value-added" metric is obtained; the Model produces these metrics.
2. The transformative quality is the delivery of power that is given to the student. This means empowering students to influence their own transformation.

   The above allows that:

   2.1. The student is involved in the decision-making process that affects its own transformation to a certain extent; the learner must take ownership of the learning process and acquire responsibility in determining the style and delivery of learning.

   2.2. In addition, the transformation process itself provides the opportunity for self-empowerment with positive consequences in the decision-making process itself.

Some ways to empower students and make them protagonists of their own transformation are: their participation in the evaluation of teaching, the selection of some curricular activities to decide whether to be elective, tutored or by directed study, the use of contracts, learning with the industry, and the development of critical thinking.

Empowering the student involves transforming the student's conceptual ability and self-awareness, which entails a degree of threat to the teacher, since it causes not only a loss of control over the structural organization of the academic activity, but also over the intellectual processes. The empowerment of students is tantamount to involving the consumer in settings standards, and that the quality is judged in terms of the democratization of the process and not just the arithmetical result.

Empowering the student, together with the notion of "PLUS or added value", approaches the definition of excellence, i.e., do things well.

5.6.4. Quality and Digital Transformation

DT seeks to improve *productivity, competitiveness,* and *quality* of the *university.* It is an accelerated evolution. It is revolution because of its radical implications on people and infrastructure, and that these demand new *educational-business* models. DT requires *globalization*, digital innovation, and to make the organization curricular and administratively a flatter one. You can still work with obsolete technologies; it does not mean getting rid of all the systems to buy them again.

5.6.4.1. Justification

They both are unprecedented *opportunities*, and they *improve the development* of academic and administrative products. Likewise, they *increase the know-how and optimize* relationships with *users*. The DT space is one of accessible technological solutions.

5.6.4.2. Requirements

1. Disruptive thinking;
2. You must start NOW;
3. Continuous training at all levels;
4. It is necessary to evaluate available digital talent;
5. It is necessary determine the Resources.

Synthesis: it is not fashion! The potential of DT is so overwhelming: there is no time for an *institution* to take the position of "wait what will happen": *digital competences* are no longer the plus: *They Are Imperative*!!!

In sum: the digital challenge is the new business culture based on digital skills. The former *informatization* was a process of evangelization of the Information Society, focused on ICT with *traditional analytics*, *rigid infrastructure*, *simple Internet* and *traditional applications*.

5.7. Quality Digital Transformation

The dilemma of getting on the train of technology does not exist: it is about *competitiveness and survival of the institution*. It focuses on *business prescriptive Analytics*, *agile and modern infrastructure*, and *staging capacity*; improves the utilization with dynamic IOT platforms and cloud-native apps; uses modern analytics: How to *generate value* through *data*? Where is it possible? At what time? And, finally, tries to analyze *all* possible data *empowering end users*.

Quality DT implies:

- Data integration, unify data-types and involve different sources.
- Develop a readable hyper-plan (layout) of *architecture*; it can be developed internally or by outsourcing contracts.
- It is internalized with the intensive use of management and *performance indicators*, ROI, conversion rates, qualified talent, condition of marginality, organic rate of repurchase (without investment in marketing), and SLA (level of service).
- Evaluate the virtualized experience of the *user*.
- It is not *reengineering* or *retro engineering*: it is *digital mapping*, extended horizontally and vertically.

5.8. Synthesis on Quality in Higher Education

Quality is a philosophical but quantifiable concept. The definitions of quality vary, and in a certain way, they reflect different perspectives of the individual society.

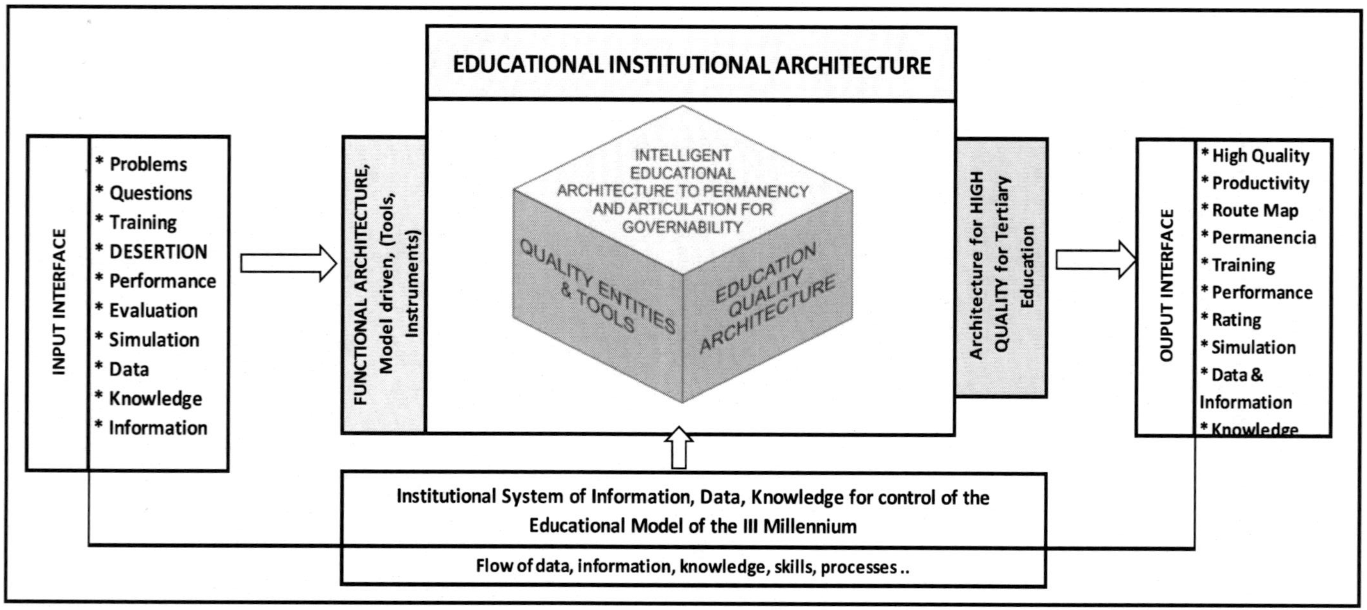

Figure 5.6. Our 3D architecture.

We presented our mature intelligent enterprise architecture – model driven, with 360° vision space –, accompanied by an executive survey on functional and non-functional structural aspects of FESSANJOSE on the road to HQ. We retrieved the best practices for HQ Accreditation that we have designed and implemented for the purpose mentioned before; also, we have developed the intelligent architecture based on knowledge and models for the competitiveness, productivity and HQ, supported in our experiences which were discussed. The architecture engine is formed by mathematical and computational models for dynamic and structural analysis of the academic governance. Computational and mathematical models are hypotheses about reality, and they allow to analyze the complex world in a more structured and coherent way. These architectures allow to follow up the progressive enhancement in search of the academic excellence.

Figure 5.6 summarizes the approach we developed in this chapter.

5.9. Intelligent Educational Architecture

Architecting implies the organization of subsystems or components to obtain new or better functionalities. If such components are knowledge based, then an intelligent behavior is obtained.

5.9.1. Space

We developed our architecture of maturity regarding the HE problem space we used to face, as displayed in the Figure 5.7. The categorization of agents of this model includes family parents, students, teachers who are the kernel, the universities, the professional unions, the industry and the Government. Students and teachers are individually represented. This architecture has allowed its knowledge mobility to several universities in Colombia.

5.9.2. The Computational Model of the Architecture

$$Y = f\ \{\text{problems, needs, performance, technological level, management level}\}$$

The results of the architecture (i.e., outputs)

$$Y: \begin{bmatrix} \text{Governability} \\ \text{Plan-Roadmap} \\ \text{High Quality} \end{bmatrix}$$

Governability: [(Teachers, Students, Resources, Continuous Improvement)]

Road Map: [(Strategic Orientation, Mission, Vision, Action Plan, Prospective Articulation)]

High Quality: [(Self-Assessment, Competence Educational Model, Pedagogical Consolidated Model, Accumulation of Intellectual Capital and Assets Innovation)].

Figure 5.7. Architecture computational model.

The computational representation is required for obvious reasons, but we need to recognize the impossibility of expressing it mathematically as a determined and closed expression. The variables types involved are objects, arrays (in terms of function coverage by level and area), vectors (technological level), fuzzy variables (e.g., performance management), and scalars. Their inputs and outputs give simple data flows to express different usually unstructured forms, e.g., a problem to solve or a plan of action to follow.

5.9.3. Architecture Computational Model

Architecture scheme could also be represented by a set of simultaneous equations given multiple inter and intra-relations of the system as iMIS inter-temporal analysis. However, the scheme described above indicates in itself, a much more rich and fertile performance to meet the demands of the corporate governability system, and is tending to suit individual conditions of an executive user. Then, computer left its condition from a simple data repository and calculator, to become an expert assistant, playing a much more important role.

5.9.4. Model for Student Intervention

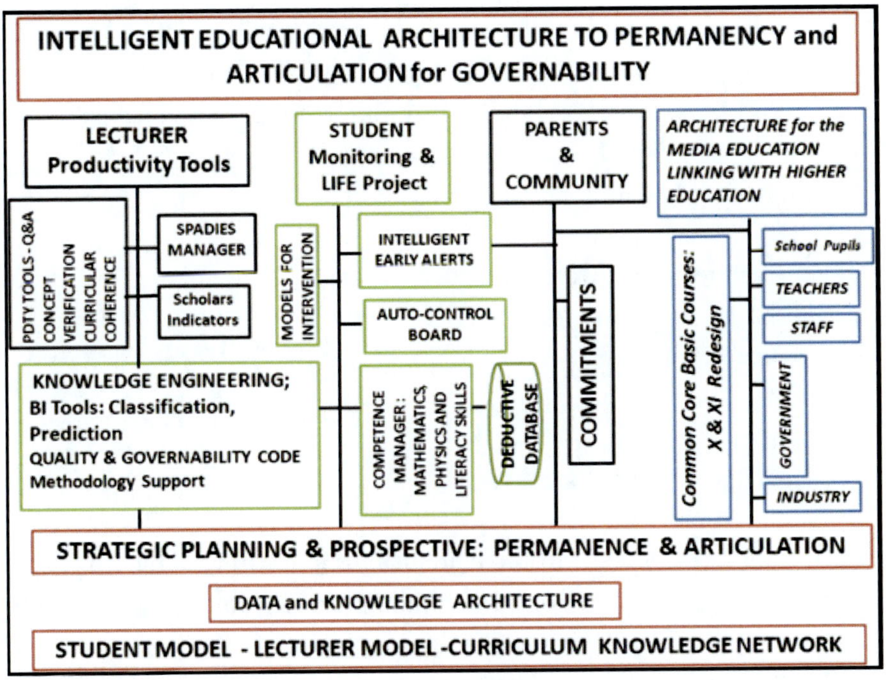

Figure 5.8. Multiagent intelligent educational architecture.

The intelligent knowledge based architecture has allowed us to define several intervention models for student remediation of several kinds: academic, psychological, economical, and related. The Intervention Model implies:

- Protocol for raising awareness: family parent's workshops for adapting to the university life, funding for HE, sexuality, etc.

- Academic difficulties protocol: study skills and time management; tests and sensitization to teachers and students about learning and cognitive styles; academic monitoring.
- Protocol for the prevention of depression: The Beck Inventory Test; identification of symptoms of depression for parents and students.
- Reinforcement workshops design on basic communication and mathematics skills to be applied to freshmen students who use to perform induction process in the first half academic period.
- Workshops design for generating skills for analysis and understanding of text, and also for basic math.
- Integral plan: training mentoring.

Our architectures were included in the Colombian Successful Experiences in Desertion "*Aprendiendo con el Bicentenario*" ["Learning with the Bicentennial"][16].

5.9.5. Lecturers and Teachers

The widespread use of different architectures as discussed in this paper provided a space to analyze and find optimal and feasible solutions, given the high complexity of the educational sector.

The teacher side includes authoring tools for content generation, and verification of concepts and knowledge.

5.9.6. Student: Early Alerts

The early alerts software for each student approaches from different fields, starting from the risk characterization of SPADIES[17] and thus, performs detailed tracking of students who are at high risk.

---

[16] MEN, 2010, www.mineducacion.gov.co,

[17] System for Dropout Prevention in Higher Education in Colombia, developed by MEN, https://www.mineducacion.gov.co/sistemasdeinformacion/1735/w3-propertyname-2895.html#.

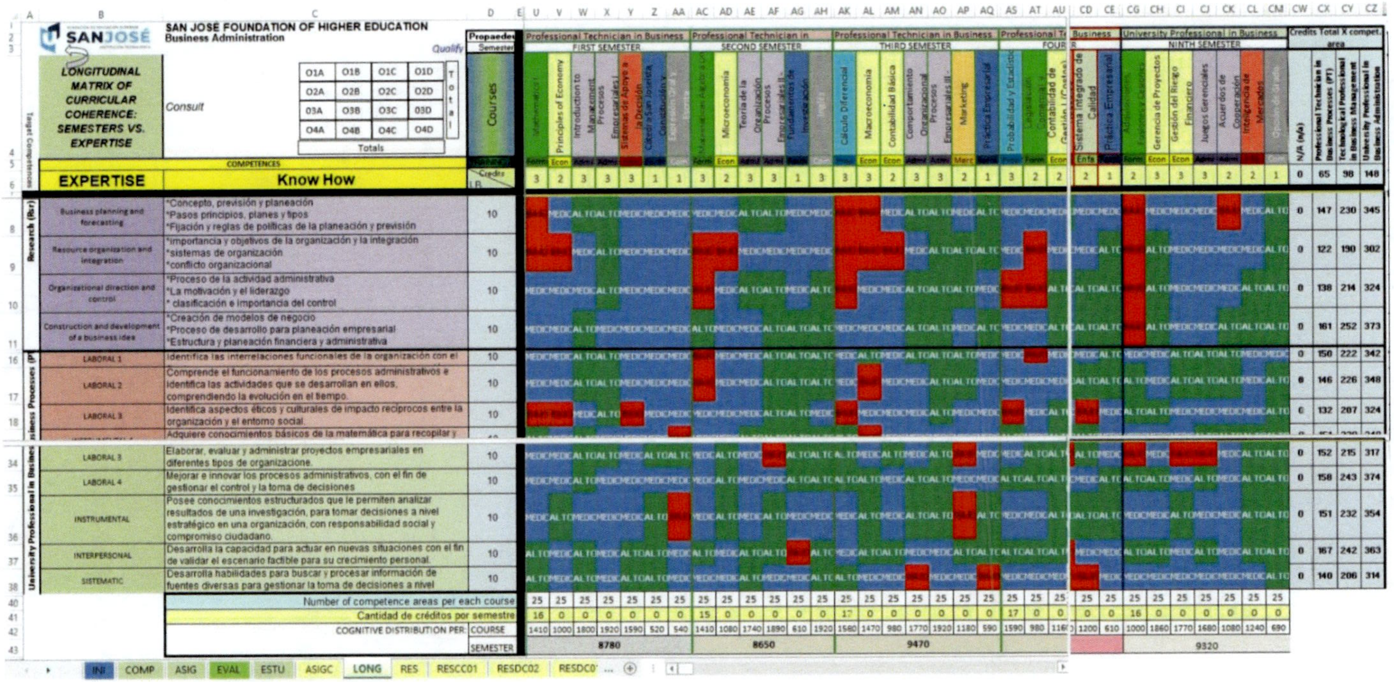

Figure 5.9. Student competence monitoring.

There is another area of focus for alerts regarding the attendance by teachers and students with a daily log of their assistance, which is carried out electronically, giving us an indication of both students and teachers who have stopped attending for a certain time, on their rate of absence, that may promote a decline in academic quality. Other item has to do with the relation of student's academic performance and their payment methods, to make also an accompaniment in this respect.

The Model monitors students by dependencies with protocols, and determines routes with online guides to improve *math, reading* and *writing* skills, including ICT-based support and protocols for the use and monitoring.

5.9.7. Student: Self-Control Board

Student Self-Control Board is a student-centered tool to monitor the students' progress and their individual information, competences, academic and administrative information. We constructed a *matrix* prototype, regarding *objective* competences vs. *micro-competences* for each subject of the curriculum.

5.9.8. Student-Teacher: Knowledge Engineering Tools

Business Intelligence (BI) tools are required for mining and knowledge discovery for classification and prediction in terms of risk and desertion. They have been important for data analysis in order to improve the models for intervention.

5.9.9. Applications and Uses

This later construct is driven to:

For *industry*:

- Find strategies that satisfy the contractors.
- On the other hand, motivates the graduates to join a company sooner.
- Identify the strengths and weaknesses of the graduates, measured by the competencies of the Global Society.
- Reduce the productive coupling time.

For *educators*:

- Makes possible to identify duplications and inconsistencies of knowledge.
- By cognitive metrics, help eliminate distortions, and, especially, balance the curriculum for optimization.

For *students*:

- Allow the students to improve their *life project*.
- Tracks students' competencies, achieved with their academic performance.

5.10. Competence Manager: ICOACH

iCOACH is an intelligent, knowledge-based tool, a high-computing instrument designed to follow up each student in the engineering faculty's academic first terms at the FESSANJOSE, specifically in the areas of *math*, *physics* and *reading-writing skills*. The software architecture has 3 main parts: Edumatic Builder (a productivity tool for the *teacher* with competence manager), the Student Model, and the Knowledge Base for concept verification. Several previous works have preceded this project, so our methodology and related knowledge are very experienced. A prototype was built to demonstrate the iCOACH validity and applicability, and also to demonstrate it as an academic productivity instrument for competence management. Below, is a scheme of our *knowledge* competence manager we are working on:

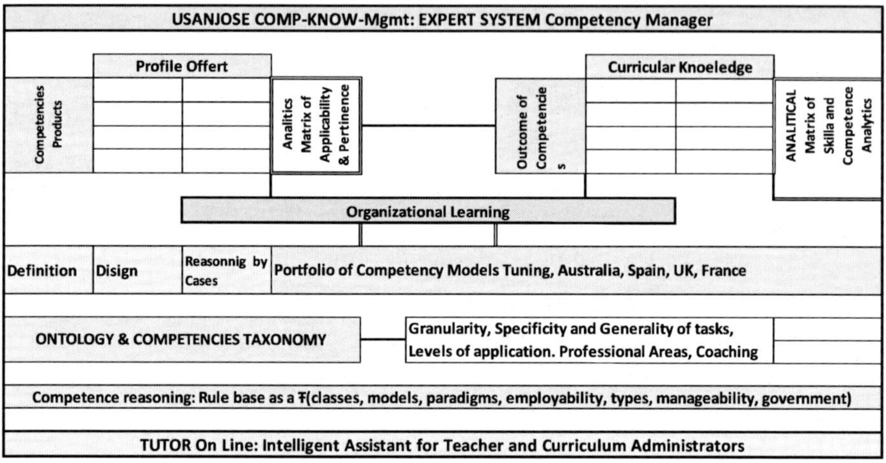

Figure 5.10. Expert system for competencies management.

The *architecture* USANJOSE COMP-KNOW-MGNT is an *expert system* as a *competencies manager*: it allows to enter needs to obtain guidelines for the teacher that help in the design of competencies and micro-competences. The final purpose is contributing to the organizational learning:

1. The Relevance Matrix allows to reason on the relations between results (profiles) and the objective-competences to obtain.
2. The Competence Coherence Analytical Matrix, makes it possible the reasoning about the relationships between target-competences and the micro-competencies of specific curricular training plans.
3. A classification of models and *competences* systems is required, through ontological reasoning, explaining *granularity*, *specificity*, *generality*, *tasks*, *levels of application*, *professional areas*, *coaching*, and others.
4. Competence Reasoning Rules System: Rules Ŧ (classes, models, schools, employability, types, manageability, government).
5. Tutor: smart assistant for the teacher and the curricular managers.

5.11. Linking ME & HE

There are various kinds of *articulation* between processes, cycles and levels of education, and between these and the Industry, the Government and the Society. The most serious problems arise in the articulation of ME with HE.

Also, there are other complications: how is the joint of educational coverage with the quality of education itself, and the articulation between the propaedeutic cycles in union with the integrative skills that are promoted, from the basic to the disciplinary.

Similarly, we could try other extensions such as: to articulate quality educational excellence, the dimensions of the productivity and the quality and competitiveness of the education sector (which requires relevant education), to reach an ongoing process as a mediation to articulate the Information Society with the building of the Knowledge Society in what leading countries are engaged.

It has been a constant and ongoing activity to improve student's skills in math, physics, and reading and writing, which allows high school attendees to enter HE showing success in curricular flexibility and a substantial improvement in their state tests results. This component allows students in high school institutions to have articulation agreement with the *university* to access 3rd semester once they have completed their ME.

5.12. Prospective and Strategic Planning: A Tool for Future Analysis

The *strategic plan* articulated with the *prospective*, allows defining the roadmap to align human, physical and technological efforts toward HE *competitiveness*, comprising SWOT (strengths, weaknesses, opportunities, threats) matrix, *critical success factors*, *strategic orientation*, *strategic map*, *prospective* and *scenarios*, *desertion*, and *addressing* ME-HE *articulation*.

A major product in this respect is a Handbook of Quality and Governance in Student Dropout and Permanence[18], including the definition of institutional policies.

5.13. Enterprise Information Architecture for HQ

5.13.1. Architecture Justification

The *architecture* is the statement towards the system determination and development, which is appropriate for the problems of HQ and competitiveness, and is the communication channel between the involved agents. It facilitates planning, management and execution for the *university's* strategic development. It can also be seen as the expression of persistent characteristics and principles that support the HE system to guide acceptable permanent changes.

It enables verification and comparison about whether the implementation of academic and administrative system is in accordance with the architectural description. Many of our innovations have been the result of a careful study and insights obtained at the architectural level.

Generally, the university enterprise architecture is composed of *strategies*, *academic* and *organizational processes*, *human* and *technology* resources, *intellectual processes*, infrastructure and other architectures. This enables the coherence and consistency among architectures.

[18] Compendio de Calidad y Gobierno en Deserción y Permanencia Estudiantil [Compendium of Quality and Government in Student Dropout and Permanence]. Convenio 755, Ministerio de Educación Nacional (MEN). Bogotá: FESSANJOSE, May 2011. ISBN 978-958-46-1442-1.

5.13.2. Information Architecture Description

About the architecture integration, there are main informational components:

- The related to business enterprise model, which functions are related to the functions that are typical of the university mission.
- The component that allows access to *institutional and strategic plans* for each *academic program*, and the *intellectual assets as production*: papers, books, software; e.g., SIFES (Sistema de Información para la Educación Superior [Information System for the Higher Education]at FESSANJOSE), etc.
- The related to the management component as support for the internal and mission processes.
- The ICT architecture: PLANESTIC, and the Virtual Self-Assessment Program of FESSANJOSE.
- The iMIS (SIFES): functionalities, their *modules* and *documentation*; the architecture for early warnings; FESSJ-ADVERTOR and LINK ME-HE (see Table 1.3); and the know-ledge mobility with other institutions in Colombia.
- The quality and self-assessment architectures, which are discussed later.
- The Documentation Repository Center: *government and internal regulations*, the whole academic programs *master documentation*, etc.
- Infrastructure: computer technology, facilities, buildings and labs.

5.14. HIGH QUALITY IN TERTIARY EDUCATION

5.14.1. The Quality Architecture

Quality has its ethos in a HEI. The accreditation of HQ is the verification completeness of the whole institutional quality, which has many different sides and is a complex problem to determine. It is a system whose components are integrated flexibly towards educational excellence. The self-diagnose of the current state, allows to determine the distance to the perfection limit. The HQ of the HE is necessary but not sufficient to achieve the sector competitiveness.

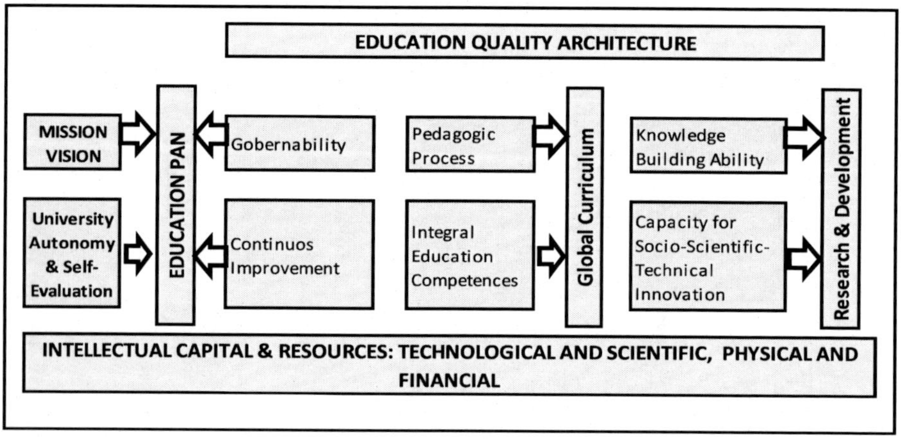

Figure 5.11. Quality architecture for HE.

These innovative developments have been the result of a long processes of submitting our programs of FESSANJOSE to the education authorities, including programs of the 3 educational levels (*professional technician*, *technologist*, and *university professional*), and of several areas (*engineering*, *business administration*, *design and arts*, and also *sports*); in total, more than 30 submitted programs in the last 6 years, with accountability of education authorities and rigorous academic peers reviews, to meet the quality standards for obtaining the Qualified Registry, as required by the MEN rules in Colombia. Also, there is the contribution of our outcomes of huge experiential learning collaborating as Academic Pair with the National Intersectional Commission for Quality Assurance in Higher Education (CONACES) and the NCA.

The *quality accreditation* (QA) has four main components, in our way of looking:

- *Educational plans* which define the *mission* and *vision* horizons; the use of the university autonomy accompanied by self-assessment; the governability that focuses on the feasibility of making the necessary decisions; and the continuous improvement.
- The global curriculum typical of the KS and the educational opening with modern educative process, which is focused on the student learning personal autonomy and on the competence-based integral education. One of the QA components is the *in-house internationalization*. In FESSANJOSE we started this process several years ago. Nowadays we have the MEN support and accompaniment, with some of the most important Colombian universities like La Salle and Rosario. And recently we have received the mentoring by representatives of the University of Yale and the Kentucky State University inside the International Mentoring Program provided by the MEN. Accreditation could be an important tool to enhance and improve cross-border education. The term *accreditation* describes the process of certification and the results of it, as mentioned before.
- Research and *innovation* to develop the capacity to produce knowledge. The knowledge architecture is discussed in Section 6.16 (bellow); it allows us to quantify the knowledge production and the accumulation of *intellectual assets*[19].

[19] Our knowledge production can be seen in www.usanjose.edu.co, *Grupo Internacional de Investigación San José EIDOS* [San José EIDOS International Research Group] – Cod. 0084828 –. We were recognized in 2010, classified and categorized in 2014, by COLCIENCIAS: Colombian National Institution of Science Technology and Innovation.

- The infrastructure refers to the required intellectual capital and knowledge assets, the financial resources, including technological ones, and the total quality from the beginning to the end in the pursuit of the excellence.

5.14.2. Governability: Knowledge for All

The integrated architectures and models for the KS promise to be a tool for analyzing and quantifying the HE *competitiveness* in this country, with the best potential for HQ by the HEI.

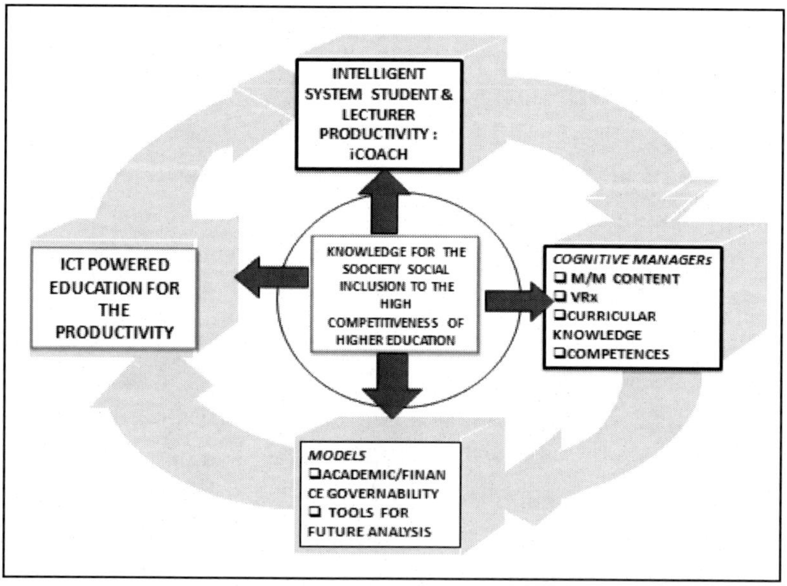

Figure 5.12. Knowledge for all.

The above diagram highlights the methodological framework for learning and prototype development, which allows us to identify problems and provide answers by the same engineering rigor, and to summarize our research efforts that propose solutions which combine intelligent virtual environments, both as computational and mathematical models for the purpose of providing systemic answers to the problems of dropout and repetition in HE.

5.14.3. Governability

It is a concept that goes beyond what an administration is, and is associated with the conditions of the institutional capacity to deliver the educational goods and services, the ease for decision making, the management of the new intellectual capital, and related, to meet the needs of government, business and society in general.

It is affected by the overload of demands and social requirements and factors such as the trust, the participation, and the consensus building. *University governance* and *financial sustainability* are closely associated. They are not a matter exclusively of educators, neither the sustainability is a purely financial issue, but finance is a strategic component of the strength of educational institutions.

A model of corporate governance is the composition of the characteristics: consistent, coherent, concerted, committed, and participatory; and is assumed by the set of systems and actors, working about ways of thinking, decision taking, acting and learning, that shows an *institution* in the different dimensions of its strategy.

Also, the governability problem is related to the mobility of students between the cycles of formation and the levels of the productive sector, as well as the establishment of a TP (*professional technician*) – T (*technologist*) education, with international competitiveness and quality.

5.15. Continuous Improvement

This architecture is not a simple conceptual and methodological improvement, but an innovation in the comprehensive and coherent treatment of the concurrent problem *desertion-permanence-repetition-quality*, in two ways: the use of creativity in the analysis of problems, and the use of creativity in solution of problems. One consequence is the possibility of a dynamic analysis, as illustrated in Figure 5.13.

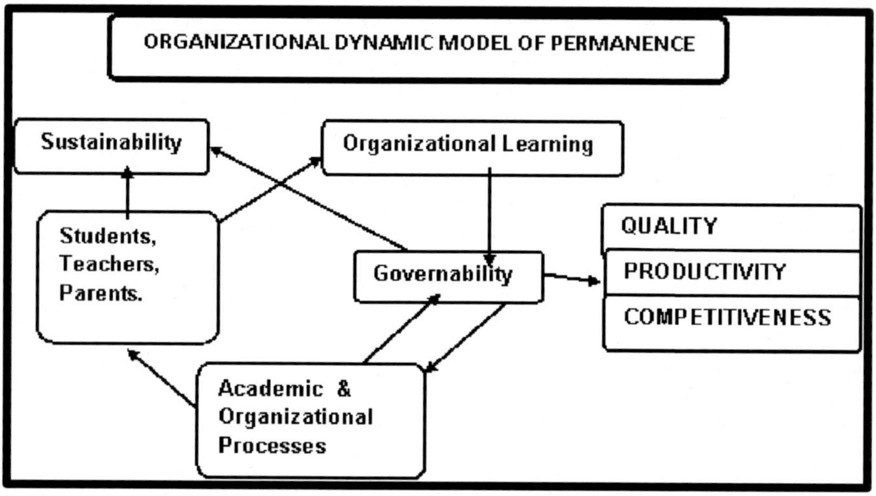

Figure 5.13. Permanence dynamic analysis.

The model focuses on the organizational governability with its impact on financial sustainability, knowledge generation, and institutional learning (starting from *young seed research groups*, up to *quality, productivity and competitiveness*). The core mission focuses on student productivity (supported by an intelligent system) and teacher productivity (supported by an intelligent handler of competencies and an intelligent courses-contents generation tool, with the system for coherence curriculum optimization).

5.15.1. Pedagogic Model and Integral Education

5.15.1.1. Pedagogic Model Description

In strategic alignment with the IEP (*institutional educational project*), the pedagogical model promotes the development of multiple intelligences to be developed during the preparatory stage of cycles, as axiological aspects of the subjects in the student *life project*.

Y= ***f***(X)

ϵ: Competence Areas

X= {
- **X**$_1$= Multiple Paradigms: social interactionism, constructivism, discovery learning Bruner, Ausubel Meaningful Learning, Learning Analytics
- X$_2$= dimensions: biological, cognitive, affective, praxeological
- X$_3$= ICT

Y: {
- **Y$_1$= Pedagogical strategies set**
- **Y$_2$= Pedagogic process strategies**

Figure 5.14. Pedagogic computational model.

It encompasses the strategy and pedagogical decisions rules which are an important component of the institutional navigation chart, the IEP, and, more specifically, of the aims and objectives of the strategic processes and action plans, focusing on skills training for evaluation and permanent control.

A pedagogical model is part of the corporate strategy, drawn on pedagogical mediation by ICT and the support of the new learning theories in virtual environments known as computational *mathetics* (from the Greek *manthenein μανθάνειν*: to learn). The *consolidated pedagogical model* promotes strategies that are required for the design and implementation of the whole pedagogical processes inside the HEI.

The FESSANJOSE's pedagogical model is configured from its constituent principles, which correspond to the following, regarding the conception of the subject (person), and the training:

- Framed within the features of Colombian Education Law.
- Established by the conceptions of the person from the perspective of the integral education.
- Principles in relation to the conception of education and the project life skills, profession and communication provided under the vision

of the document on the four pillars and the seven lessons for the future education of Morin[20], are.

- Stated from the perspective of social inclusion, human development and welfare.
- In connection to educational and pedagogical concepts: the developing skills to generate meta compressions and understandings from the reflections.
- From the question as education strategy.
- Through the significance of meaningful learning experience.
- Through interaction in teaching.
- From the perspective of the persons' intelligence under the multiple intelligences look.
- Through understanding the pedagogical models.
- Located epistemologically in the constructivist model.
- Through strategies and active classroom participants.
- From the problem solving. From projects generation that facilitate learning through the acquisition of knowledge and problem-solving situations.
- Through the interactions of the subjects (teachers and students), the relationships established, and recognitions achieved.

5.16. Knowledge Architecture

5.16.1. Conceptualizing

The knowledge architecture, can be defined as the set of strategic design decisions that affect the structure, intelligent behavior or functionality of the system as a whole. The HEI are the natural space where the ethos of knowledge, can be developed and the *intellectual capital* (IK) can be accumulated. It is no secret the inadequacy of the iMIS to incorporate

[20] Morin, Edgar. *Seven complex lessons in education for the future*. Paris: UNESCO, 1999. [Onlilne] http://unesdoc.unesco.org/images/0011/001177/117740eo.pdf.

knowledge. In fact, is shared by many authors that well over half the market value is explained by the IK in companies of the Third Millennium. Traditional systems do not account for the IK and only estimate the in-book values vs. market values, when selling is focused on negotiation processes. On the other hand, there are already quite a few financial models of the knowledge economy that consider the IK. For the universities this is a critical issue in the way of the HE competitiveness.

5.16.2. Knowledge Process

A process is defined as an activity that takes place in ongoing stages to produce a result or a specific group of results, coherently. It is a set of activities that transform inputs into a product or a service with customer value.

A knowledge process is a sequence of iterative activities characterized by having measurable inputs/outputs and value-added activities, including the organization, staff, equipment, procedures and materials required to produce a result: knowledge, intellectual assets and intelligent performances.

5.16.3. Innovation or Innovation ICT?

An unanimous consideration exists on that practically there has been no technical or scientific change that has not had an impact on pedagogy. The education sector has been profiled in two ways, sometimes contradictory: (1) a changing and modifying way, since the education sector must educate future generations with the scientific, social and technological changes; and (2) the conservative way, given that education must preserve the social and cultural structures, which it should favors.

But it is also noted that, behind, there are the benefits of such changes. In these days of rapid obsolescence is when the education sector will require more innovation, and take advantage where possible from the developments in other sectors, besides, enabling the transformation of the knowledge appropriation practices from different compartments of science and a variety of cultures.

The question remains as whether the innovative activity in computer science education, the set of social research, should be the progress of scientific basis, or else, its origins should be in the progress of the computers according to the possibilities of their profit to be contemplated.

5.17. Mathematical Knowledge Architecture

The Mathematical Knowledge Architecture comprises the Leontief *input-output* matrix (innovated to the KS) in a system either at the micro level (i.e., a program of study) or at a higher level of an institution or a country. This architecture has 3 components:

- The inter-sector industry subset, including the *knowledge sector.*
- The restriction and threshold values: real world facts.
- Various objective functions.

The model was integrated with linear programming to analyze together the *education* services in the *tertiary sector* of the economy (trade and services), and the primary and manufacturer sector.

The optimization criteria are shown schematically in Figure 5.15 which shows the levels of students in HE related to permanence/desertion.

The *governability* deals with the control of the flows within these levels of students, outlined and quantified in the multiple objective functions. It seeks to maximize the number of students performing well given the facts of the HEI's world (restrictions, limits), which is one of the indicators of quality pre-excellence. The academic governance also implies the control flows that are contrary to policy optimization. The model also generates information on the flows of student's costs.

$$f(X, D_f) = 0 \qquad (1)$$

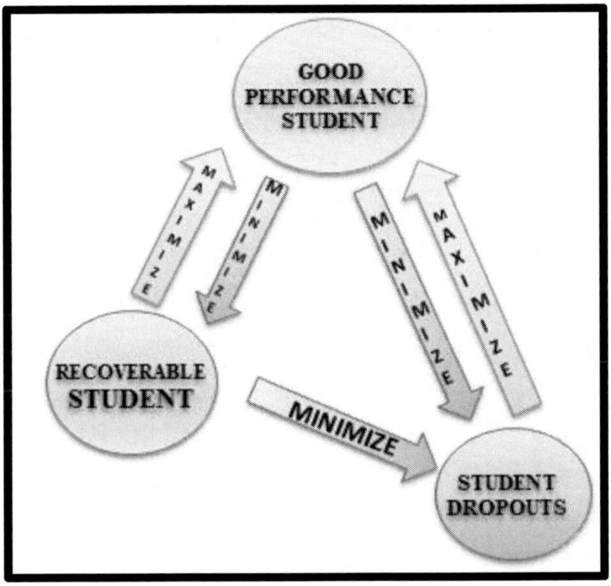

Figure 5.15. Student levels for optimization.

D_f is the final demand. Including *households*, *inversion*, *government*, Δ *inventories*, *intellectual capital*, *rest of the world …*

Political function issues: (2)

Max Z= $f(\Delta Students\ Level)$(2)

$$(I - A)X - D_f = 0 \quad (3)$$

F (resources)≤ b_i availability (4)

CONCLUSION

The High Quality architectural methodology illustrated us how the HE complex system can be composed and articulated in order to show how is the governance and its relationships to actors, major subsystems, and software, to obtain new, important and better functionalities. This architecture conforms a big system as an organized and complex whole.

The architectural approach allowed us to analyze and evaluate large volume of academic and administrative information, in a well-structured way, with 360° view. It also allowed us to browse the entire administrative and academic information of the *university* (walking-through from top down to bottom up).

Greatly, the applied method facilitated the strategic alignment of the different units to the final strategy: *the HQ*. Also, it provided the synchronization of academic and administrative processes, establishing common referents for the whole community.

The congregation of information at different levels of aggregations reduced the complexity of the analysis and design, as even facilitated institutional improvement of plans coherently.

The *self-assessment* itself, managed as an organizational process with metrics, facilitated the linear programming and the Leontief input-output models, extending their value for institutional optimization purposes.

Another important contribution of the Model of Self-Assessment for HE is in accordance to the NCA, for which we use some tools for *big data* management and visualization through a multivariate fuzzy matrix–dashboard with a 360° *architectural vision*.

With the *self-assessment* as a process variable, i.e., it was allowed quantifying the incidence of *self-assessment* in terms of quality, academic sustainability, student desertion, and many other factors, thanks to the knowledge architecture.

Finally, we consider that the information enterprise architecture is the largely expected *institutional memory*.

Synthesis: Outcomes of the Architecture

The analytics of the university product Self-evaluation and High Education Quality which is supported by the IA (using an FDM), is another tool that allows us to permanently evaluate the progress of our DT, what nourishes us (feedback) and guides us with the anticipatory information of the future (feedforward).

The *accreditation of high quality* is the verification of the completeness of the whole *institutional quality* towards the excellence: our architectures help us visualize, diagnose and quantify as we are approaching to academic excellence.

These experiences have been cataloged by the MEN as successful and this has allowed us the mobility of knowledge that we are implementing in several universities in Colombia.

Regarding that methods of HE are changing rapidly, we are facing new challenges both at institutional and at individual levels. Due to the increase in national and international competition, institutions are struggling to offer attractive academic programs with HQ Accreditation for learners by employing modern ICT based e-learning.

These architectures allow us to follow up the progressive enhancement in search of the academic excellence.

Chapter 6

Intelligent Architecture of Position Audit of Digital Transformation USANJOSE-Arch-Level-DT

Abstract

The digital interruption is driving the transcendental transformation in the company that responds to the risks and threats of digital disruption. It must integrate digital capabilities into the heart of the business, making digital literacy a core competence, and not a particular circumstance stubbornness.

The audit, etymologically speaking, comes from the Latin verb *audire*, which means 'to hear', which in turn has its origin in the first auditors who exercised their function judging the truth or falsity of what was subjected to their verification, mainly observing. However, it is also said to come from the verb in English *to audit*, which means 'revise', 'intervene' or 'evaluate'.

Generally, audit is the action of verifying that a certain event or circumstance occurs institutionally, according to a plan, but if the audit is discussed in an organization, it refers to the tests that are carried out on digital, operational, cultural information, budgetary, and/or organizational based on the execution of the strategically planned actions, as well as the policies and guidelines established by the entity itself.

Keywords: classification model of digital maturity, position audit, classification items

6.1. AUDITING THE INSTITUTIONAL POSITION

6.1.1. The Need for Evaluation of a System

There are numerous studies that evaluate the quality of a system, focused on measuring the quality characteristics exhibited by the system *today* vs. the excellence, e.g., the functionalities, such as:

- System response times, etc.: allow knowing if the system performs the processes in an appropriate way.
- Quality of data, knowledge and information: refers to the quality of the outputs that the system offers.
- Use (usability): refers to enable the user to operate the system and be a beneficiary of the same technology. It is associated with the measure of the level of satisfaction from the side of the users who use the system in the organization: e.g., its contribution to the productivity of the *student* and of the *teacher*, that is to say the *individual impact*. It represents the level of influence that the system exerts on the processes that the users carry out and its contribution to the productivity of these users.
- Institutional impact in systemic terms of the iMIS mentioned above. HQ *educational productivity* and *competitiveness*.

6.1.2. Current Digital Maturity Leading to the DT

We present a hypothesis of levels of digitalization in *universities* on the way to the maturity of DT. We rely on different experiences of important research firms and structured interviews to companies in the world, but especially on our own experience for many years working in HE *research centers* and with public and private companies for which we have been hired as consultants in *information systems* issues. We used the Nolan Model at that time, and also directed several *postgraduate theses* and *pre-grade projects*.

In relation to the aforementioned researcher Richard Nolan, he developed the multistage model for *information systems* of the last century interviewing hundreds of executives of companies in the world, and he left us many qualifying experiences about the iMIS.

6.1.3. Analysis of Stages for the Nolan Model

In summary, the phases consist as follows:

- Initiation stage: this phase is part of the implementation process since it has begun to introduce all the information systems within an organization.
- Stage of contagion: this phase is part of the post-implementation process, since the information systems are already used in a more disseminated form and are accepted by the majority of departments within the organization.
- Stage of control: this phase is also part of the post-implementation period, since once the system is fully applied, users begin to feel frustrated with the system, as this does not work according to their expectations, presenting delays, failures, largely by the so-called software crisis, among others.
- Integration stage: part of the post-implementation phase in which there is total acceptance of the systems by users who demand new features, in order to better perform the organizational processes.
- Administration stage: part of the post-implementation period also, in which mechanisms are created to manage the systems and perform the maintenance thereof, to try to keep them up to the top of the wave.

- Stage of maturity: this stage is part of the post-implementation phase since users feel complete confidence in their use of the systems for various organizational processes.

The methodology consist in that, once the level of progress/evolution of the iMIS is characterized, it is said that, with the assumption of sequencing the phases without skipping any of those stages, the strategies to follow in the journey to maturity are easily inferred.

6.1.4. For the Purpose of the DT

We will concentrate on 3 scenarios or *loci:*

- *Culture* and *digital talent*: a university's approach to digitally driven innovation, and how to empower stakeholders and employees (teachers, staff) with digital engineering.
- *Organization*: the alignment to support the digital strategy, the governability and the execution of the DT.
- In addition, the *technology*: use and adoption of an emerging and exponential technology.

Obviously, the reference is our DT Ecosystem that we have identified, from which we will take the other elements and factors.

Audit is a method of research and analysis, that aims to make a thorough review and evaluation of the several scenarios mentioned, critically and systematically examined by a group of qualified people (experts) independent of the audited institution, where the main purpose is to express an autonomous and competent opinion about the state of digital evolution to achieve the DT, supported by the evidence of internal policies of digital culture and governability, budgeting, investments in technology, the organizational level, empowerment of the digital talent, operational excellence, and expectations for the future.

6.1.5. Our Characterization of the Long Journey towards the Maturity of DT

We have characterized 3 levels in this way:

6.1.5.1. Level 1: Initial Level

The initial ones are institutions which invest little in exponential technologies, with very low budget:

- They are biased towards traditional services, with very low administrative and administrative productivity.
- Have limited capacity to appropriate and internalize *knowledge*.
- They have limited use of online offer channels (social media).
- They run few digital academic programs with less marketing programs (which also do not prioritize).
- Qualified users are absent, with poor customer service culture strategy.
- The organizational strategy is unclear towards the DT.
- These *initiates* spend budget hiring the wrong talent, away from the required digital talent.
- It is the stage of digital initiation: just beginning a long DT journey that requires attitude and willingness.
- The information systems simply offer the daily needs but do not explain satisfactorily the outcomes, and their contribution to the productivity is not enough.

6.1.5.2. Level 2: Implementation

Institutions in this level concern about investing in technology with improvements in digital policies and strategies. Characteristics:

- They go beyond from the simple concern for *the digital.*
- They are willing to invest in extreme architectures (e.g., *big data*, CRM, e-commerce) that they need to scale to their interest towards *the digital*.
- Emphasis is on the supply side of products and services, but little attention is paid on customer demands.
- Small digital marketing teams to create informed customers.
- Important improvements in academic digitalization and in productivity.
- Advances in administrative digitalization, but low rates in competitiveness.
- Upward data strategy but does not include the perspectives of users and suffers from security problems.
- It is the stage of DT implementation: much effervescence, but relatively little concreteness.
- The outcome of the information systems appears but its contribution to productivity is incipient and limited.

6.1.5.3. Level 3: High Degree of Maturity

In this level, the *digital engineering* leverages academy and administration with institutional competitiveness. These institutions:

- Are more competent to collaborate internally and externally to allow practice (knowledge) and academic-administrative innovation with digital engineering.
- Emphasize the general experience of the informed customer.
- Have *strong* coordination and continuous communication between digital marketing and the DT.
- Make the distinction in the competitive environment, making use of creativity and analytical rationality with models and architectures.

- Are better in terms of high academic quality, high quality of information, and high quality of curricular and administrative processes.
- Use predictive analytics (machine learning) to know and measure the experience of the informed client and the scope of their academic-operative processes for the important generation of income.
- Like the most mature universities in DT, use technology to combine marketing, content, and transactional contact points in cohesive digital experiences, e. g., customer satisfaction.
- Extract (infer) the satisfaction of the informed user to obtain information about the emotional signals that trigger the acquisition (i.e., physical and digital world combination); this level is the *post-implementation* stage: DT administration.
- Have information systems that explain the final results, and their contribution to the productivity is important in the whole organization.

6.2. INSTRUMENTS OF THE POSITION AUDIT

In accordance with the above, and regarding our model of DT *ecosystem*, we synthesize a FD matrix, with which we represent the fuzzy knowledge base that interrelates what we call:

a. *Organizers and previous references* $\boldsymbol{O_i^k}$(rows, i) following the characterization of each of the 3 phases of DT, with an associated *indicator* $\boldsymbol{I_i^k}$ (in order to provide institutional information to the *expert*).

b. The K dimensional *locus* or the dimensional scenarios (columns) of:
   - *Culture* and *digital talent*: a university's approach to digitally driven innovation, and how to empower stakeholders and employees (teachers, staff, family parents and other users) with digital engineering.
   - *Organization*: the alignment to support the digital strategy, the governability and the execution of the DT.
   - *Technology* and *infrastructure*: the adoption and use of emerging technologies.

Each *locus* includes 3 columns (e) to be filled by each of the 3 DT *experts* for their fuzzy *expert assessment*, which is complemented by the columns (j) of their *expectations* of future ***costs*** of progress towards the next stage of DT, the degree of ***complexity*** in the future implementation, with the expected ***productivity*** and with the *possibility* of ***income*** of the next phase.

It should be noted that each of the experts must indicate their level of knowledge e_p about the institution.

The row (c values) for *factors weights*, are those related to the values for *defuzzification* purposes.

For each k° LOCUS

$$\sum_i a_{i,j} n_j = b_i$$

$$\sum_i a_{i,j} n_j = d_j$$

Figure 6.1. Matrix balance.

The Matrix Scheme of the instruments is displayed in the figure below:

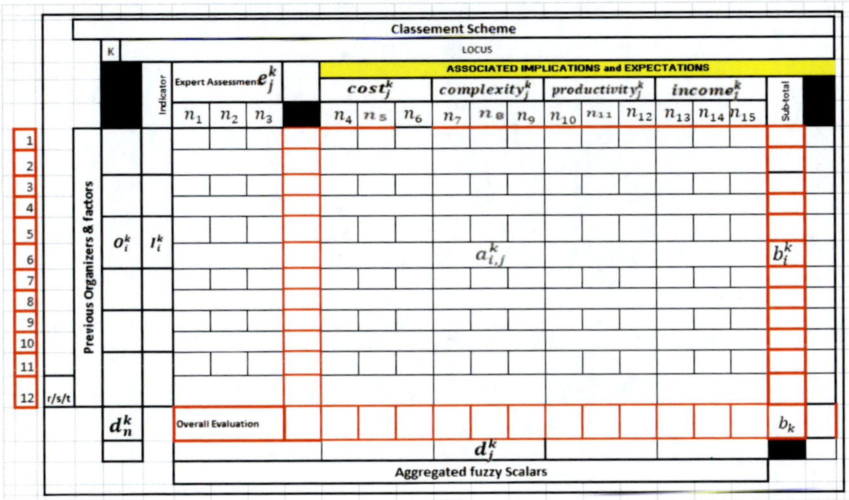

Figure 6.2. The 3D scheme of the instruments.

The 3D scheme of the instruments is displayed in the figure below:

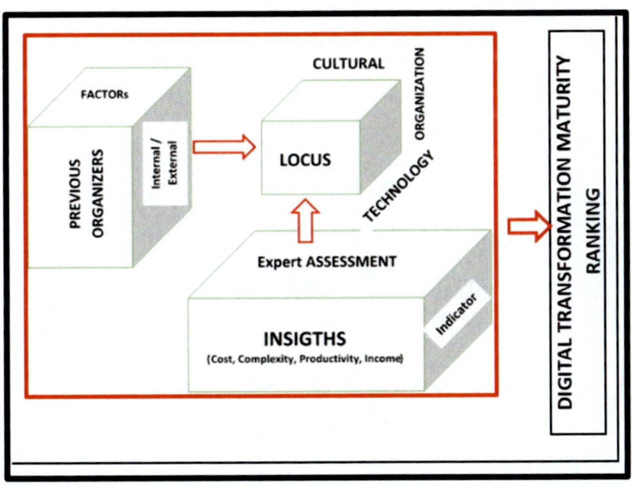

Figure 6.3. DT maturity ranking.

The 3 instrumented matrices are displayed in the figure bellow: here, the 3 *evaluation sheets* corresponding to 3 *loci* are displayed: *culture*, *organization* and *technology*.

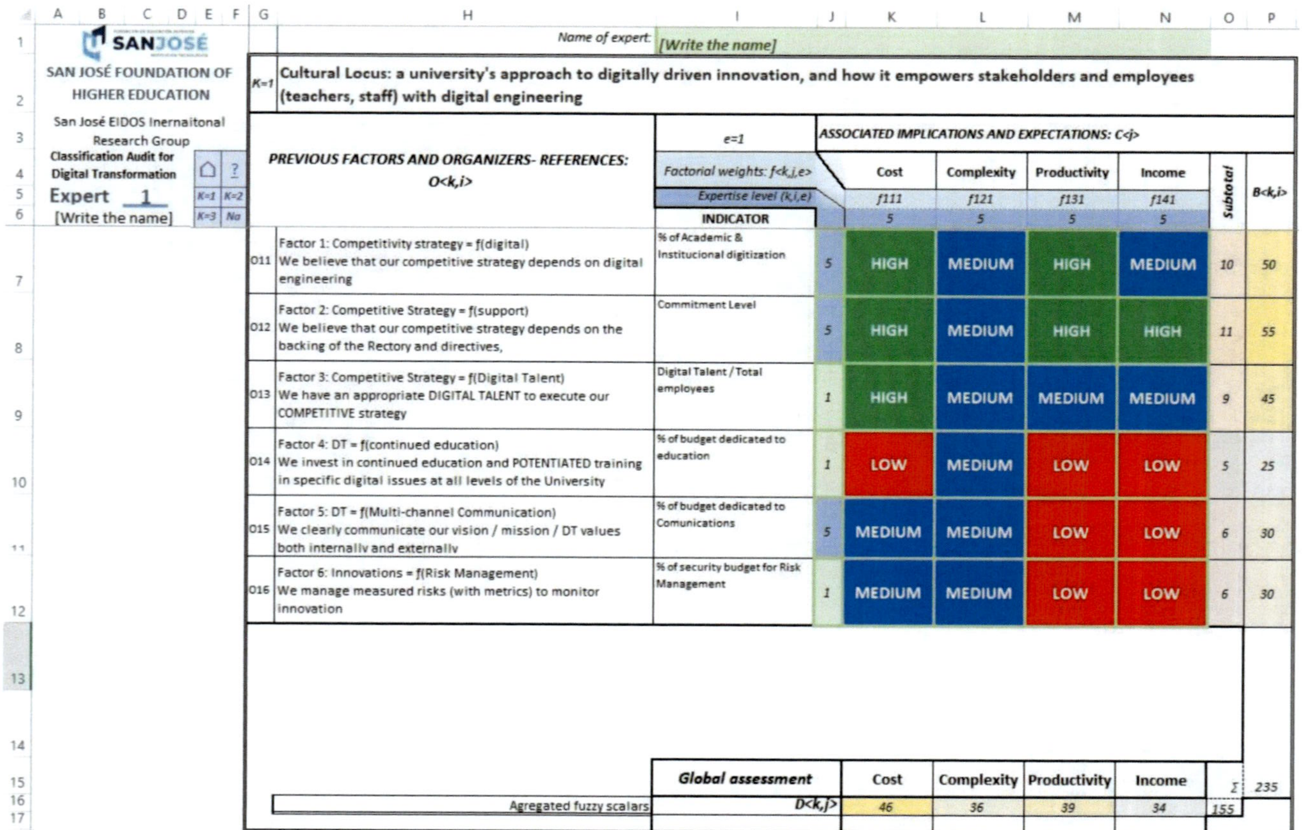

Figure 6.4. Culture locus 1.

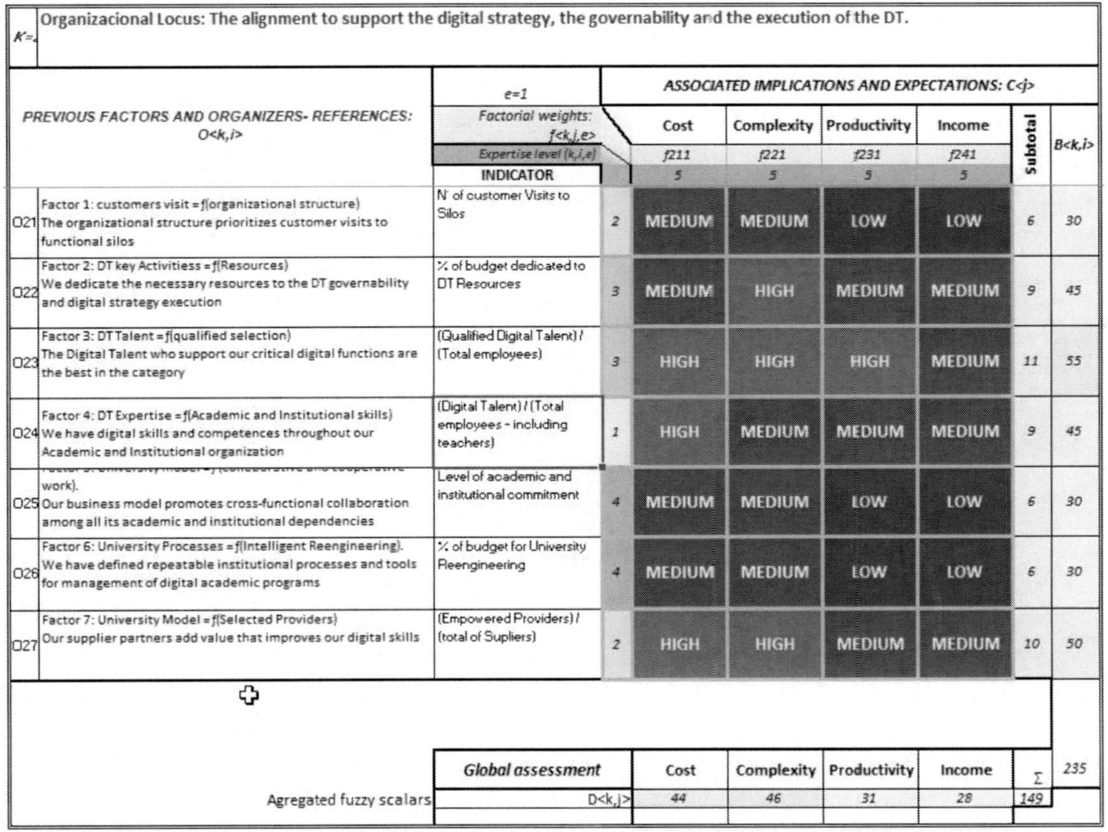

K=	Organizacional Locus: The alignment to support the digital strategy, the governability and the execution of the DT.									
	PREVIOUS FACTORS AND ORGANIZERS- REFERENCES: O<k,i>	e=1 Factorial weights: f<k,j,e> Expertise level (k,i,e)		ASSOCIATED IMPLICATIONS AND EXPECTATIONS: C<j>					
				Cost	Complexity	Productivity	Income	Subtotal	B<k,i>
				f211	f221	f231	f241		
		INDICATOR		5	5	5	5		
O21	Factor 1: customers visit = f(organizational structure) The organizational structure prioritizes customer visits to functional silos	N' of customer Visits to Silos	2	MEDIUM	MEDIUM	LOW	LOW	6	30
O22	Factor 2: DT key Activitiess = f(Resources) We dedicate the necessary resources to the DT governability and digital strategy execution	% of budget dedicated to DT Resources	3	MEDIUM	HIGH	MEDIUM	MEDIUM	9	45
O23	Factor 3: DT Talent = f(qualified selection) The Digital Talent who support our critical digital functions are the best in the category	(Qualified Digital Talent) / (Total employees)	3	HIGH	HIGH	HIGH	MEDIUM	11	55
O24	Factor 4: DT Expertise = f(Academic and Institutional skills) We have digital skills and competences throughout our Academic and Institutional organization	(Digital Talent) / (Total employees - including teachers)	1	HIGH	MEDIUM	MEDIUM	MEDIUM	9	45
O25	work). Our business model promotes cross-functional collaboration among all its academic and institutional dependencies	Level of academic and institutional commitment	4	MEDIUM	MEDIUM	LOW	LOW	6	30
O26	Factor 6: University Processes = f(Intelligent Reengineering). We have defined repeatable institutional processes and tools for management of digital academic programs	% of budget for University Reengineering	4	MEDIUM	MEDIUM	LOW	LOW	6	30
O27	Factor 7: University Model = f(Selected Providers) Our supplier partners add value that improves our digital skills	(Empowered Providers) / (total of Supliers)	2	HIGH	HIGH	MEDIUM	MEDIUM	10	50
		Global assessment		Cost	Complexity	Productivity	Income	Σ	235
	Agregated fuzzy scalars	D<k,j>		44	46	31	28	149	

Figure 6.5. Organizational locus 2.

K=3	Technological Locus: the use and adoption of an emerging technology is considered including infrastructure.								
	PREVIOUS FACTORS AND ORGANIZERS- REFERENCES: O<k,i>	e=1 Factorial weights: f<k,j,e>	Expertise level (k,i,e)	ASSOCIATED IMPLICATIONS AND EXPECTATIONS: C<j>				Subtotal	B<k,i>
				Cost	Complexity	Productivity	Income		
				f311	f321	f331	f341		
		INDICATOR		5	5	5	5		
O31	Factor 1: Technology Budget *f*(Total Budget) Our technology budget is fluid to allow the priorities of the change	(Technology Budget) /(Total Budget)	5	HIGH	MEDIUM	MEDIUM	MEDIUM	9	45
O32	Factor 2: DT Road Map = *f*(Marketing, Technology) Our marketing and technology teams work collaboratively to design our DT roadmap	Road Map maturity level	3	MEDIUM	MEDIUM	HIGH	MEDIUM	9	45
O33	Factor 3: DT development = *f*(Methodology) We have a flexible, iterative and collaborative method for technological development	Methodological maturity level	3	MEDIUM	MEDIUM	HIGH	MEDIUM	9	45
O34	Factor 4: DT Performance = *f* (architectures, technologies) We take advantage of modern architectures (IOT, Big Data, AI, Cloud, etc.) to promote speed and flexibility	N° of exponential technologies	1	HIGH	HIGH	HIGH	HIGH	12	60
O35	Factor 5: DT performance = *f* (Digital Engineering) Using digital tools which provide metrics to promote innovation, collaboration and employee mobility	% (digital engineering) / (total of others oganizational engineering)	4	MEDIUM	MEDIUM	MEDIUM	MEDIUM	8	40
O36	Factor 6: DT Excellence = *f* (External Consulting) External Consulting Contracts are made in the pursuit of excellence in DT	% (External Consulting Budget) / (Total Consulting Budget)	4	HIGH	MEDIUM	HIGH	MEDIUM	10	50
O37	Factor 7: High Quality = *f* (iMIS maturity) The current level of the MIS (internal development software, packages, or contracted services) contribute significantly to the high quality of the organizational and academic processes	iMIS level	2	HIGH	MEDIUM	HIGH	MEDIUM	10	50
O38	Factor 8: Productivity = *f* (iMIS maturity) The academic and employee productivity (including teachers) is explained by the maturity of the iMIS	{academc and institutuional productivity}	2	HIGH	HIGH	LOW	LOW	8	40
		Global assessment		Cost	Complexity	Productivity	Income	Σ	375
	Agregated fuzzy scalars	D<k,j>		62	51	59	47	219	

Figure 6.6. Technology locus 3.

CONCLUSION

For us, it is a great satisfaction, to be able to offer the instrument that will allow to the institutions of the PSEd to measure and evaluate the level of maturity of their transition from the DT to the IV Industrial Revolution.

We support our long-standing experiences with *management information systems*, for which we use various instruments and guides for maturity classification, but especially the *Model of R. Nolan*.

We are confident that this tool can be adapted to other companies.

We reviewed several classification or position audit models and our own experiences in the area of management information systems with R. Nolan models.

Chapter 7

DIGITALIZATION OF THE SOCIAL PROJECTION AND RELATIONSHIP WITH THE EXTERNAL SECTOR

7.1. INTRODUCTION

The social projection of the university is one of the fundamental pillars of our DT model. It requires the mapping of products, services, knowledge, consultancies and so on, on society, industry, government and even the rest of the world. In synthesis it refers to external relations, which obviously includes internationalization.

The interaction with the environment is the means by which *Fundación de Educación Superior San José* establishes appropriate and externalized relationships with the social, local, regional, national and international contexts. It promotes the exchange/mobility of digital talent, knowledge and experiences with the productive business sector and the society in general in the global context, especially if it's taking into account the imminent advent of the IV Industrial Revolution and the maturity of the Knowledge Society.

This relationship with the external sector, also understood in DT as the social projection (SP), being of the substantive functions to be developed in the HEIs, is assumed as an integral part of the responsibility that belongs to those who are decisional actors and work in it.

This dynamics is developed in the different fields of knowledge and in the professional-technical, technological, and professional propaedeutic formation modalities, as can be seen in the graph, which describes the architecture, Figure 7.1.

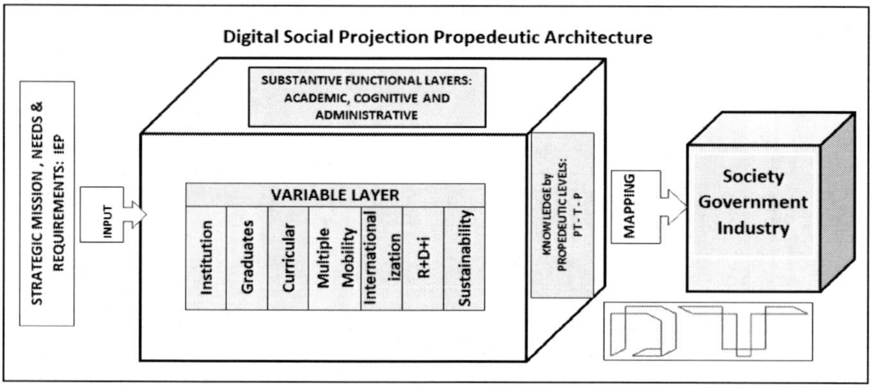

Figure 7.1. SP architecture.

7.2. Computational Model

The corresponding computational model can be expressed as:

Y = $f_{mapping}$ (Instittution, Graduates, Curriculum, Mobility, Internationalization, RDi, Sustainabiliy)
Where:
Y: Endogenization of Products and SVCs of the Insertion of the University in the multidimensional environment.
f**:** Functional structure of exogenous control variables.
Inputs: STRATEGIC MISSION, NEEDS and REQUIREMENTS of the IEP.

The above points out the interaction with the community, the mechanism between academia and society, industries, government and companies, for the transformation of the regional and global environment, as well as contributing to the sustainable development of society.

7.3. The Central Objective of the SP IS

To structure the necessary means and conditions to:

- Strengthen the institutional relations.
- The conception and execution of projects, programs of social and technical transformation, plans, strategies and methods of Problem Solving, under which the academic community impacts or projects (mapping) to the local and global society.

7.3.1. Among the Specific Objectives Are, to:

- Establish bilateral relations or business agreements, internships, multilateral programs.
- Contract on joint scientific-technological, artistic, cultural and sports projects.
- Conduct Scientific-Technical, Cultural, Artistic, etc. Conferences and meetings. Towards the fertilization of mutual knowledge and business in the networking space.
- Conduct studies of the business sector with the purpose of knowing the needs in terms of required academic training, cooperation with educational institutions and business organizations, national and international, that allow the strengthening of curricula.
- Prepare diagnoses of the weaknesses of the internal and external organizational environments.
- Promote the curricular update to give pertinence/relevance to the contents taught in the space of internationalized competences.
- Develop cooperation programs with educational institutions and scientific technical (such as the IEEE), business, national and international organizations, that allow the strengthening and universalization of the local study plans so that the graduate compete with property in the global society.

7.3.2. Special Mention to Entrepreneurship

As a way of thinking and acting oriented towards:

- Creation of wealth.
- Generate value.
- Promote the economic benefit of those involved and of the society in general, with Productivity and Quality towards high competitiveness.
- Encourage in students the entrepreneurial spirit through the establishment of their own company, making use of the modern ICT and extreme technologies such as artificial intelligence, machine learning, Big Data, etc.
- Promote in the students and graduates the affiliation to International Scientific Societies.
- Develop the culture of entrepreneurship with the implementation of contextualized plans by the University.

7.4. Methodological Life Cycle of the SP

The methodology with which it is intended (which each of the programs offered) to positively impact the business environment.

The methodological aspects summary a plan to solve problems.

- Long-Term Strategic Vision, integrated into the IEP.
- Obtain contextualized results, with integral and sustainable human development.
- Improve products, services and extensions, adjusting them permanently, to maintain their validity, and socialize them with faculty and university bodies, acting as feedback and feedforward of studies of relevance and curricular renewal.
- Support Employability and Permanent Education.
- Develop new competencies for the global renewal and deepening of curricular knowledge, including the appropriation of new digital skills.

7.5. INTERNATIONALIZATION OF THE HE

7.5.1. Background

From the Strategic Plan of the FESSJ 2006 – 2012, included from the beginning (2006), in numeral 2.3 the INTERNATIONALIZATION of the FESSANJOSE, as well as Modernization and Globalization to face the change; Faced with the institutional decision by the propaedeutic cycles, for which the Qualified Registries of the Professional Technical careers and of the new TECHNOLOGY programs were achieved, the MISSION and VISION of both the Institution and the different accredited programs were reviewed. For this purpose, the ACADEMIC COUNCIL was entrusted with the discussion and analysis of the complementary strategic elements. Presence in world-class events: numerous Papers, books and book chapters in world-class events: USA, Madrid, Mexico, Kuala Lumpur, Kuwait, Argentina, Brazil, Africa, etc. evaluated by International Pairs of the IEEE ACM ISTEC.

7.5.2. Basic Aspects

Internationalization, in accordance with what has been learned by the FESSANJOSE, should be projected as a strategy to generate intercultural processes within the institution, foster relations of cooperation and mutual collaboration, as well as the execution of multiparty projects of technological, cultural, sports or any kind other nature, involving the participation, intervention and mobility and exchange of knowledge, teachers, students and researchers of the participating organizations extending the benefits of technology transfer and its internalization with enhanced appropriation.

7.5.3. General Objective of the Internationalization of the HE

Support the process of modernization of FESSANJOSÉ in the face of globalization and globalization, in such a way that it is recognized as an HEI that contributes to the technical, economic, social and cultural transformations that the environment demands and especially the arrival of the IV Industrial Revolution.

7.5.4. Specific Objectives

- Strengthen the supply and capacity of digital talent of FESSANJOSÉ around internationalization pursuing high quality, high productivity and desirable competitiveness in the global society.
- Adapt the processes of governance and internal management with organizational learning in order to coordinate activities in a clear and transparent manner for the flowering of products, services and culture of internationalization, obviously including the culture of the pedagogical service to students.
- Adapt the curricula in such a way that the offer of programs and the training of human talent are relevant and in accordance with national and international trends.
- Financially and logistically favor all the academic, SP and research and innovation activities that FESSANJOSÉ improves its competitive position in the national and international context.

7.6. Umbilical Cords to Improve the Life Project

7.6.1. Using Digital Learning (Learning Analytics) to Develop

- *Permanent Education Program-Graduates:*

  Continuing education is designed to expand and strengthen their skills, supporting them in their professional training and improving their employment and/or their own company, updating themselves in their specific knowledge.

- *Trainings and Seminars:*

  The training is developed according to the trends of science and technology in postmodernity also according to needs based on the demand of the governmental, industrial and business environment.

7.6.2. Key Aspects of Internationalization: Mobility

It has to do with People, Intellectual and Physical Assets, Services, Ideas, Values, Online and Physical Institutions.

7.6.2.1. Academic Mobility

International student mobility (participation of a student in the university, in academic activities in another institution or foreign organization); Undergraduate and postgraduate academic exchanges, academic internships abroad, especially language courses abroad; Foreign students in the Campus.

7.6.2.2. Mobility of Teachers

(A teacher of an institution performs some type of academic activity).

- Dictating courses or seminars (guest teacher).
- Presenting international papers.
- Participation in an event or organization outside of your country.
- Making visits and research internships.

7.6.2.3. What Is Achieved?

Digital talent more modern and global, which projects great influence on the internationalization of the HE.

7.6.2.4. The Internationalization Process

- It is consolidated as a transformation process in which the activity areas of an institution are constantly adequate (strategic accommodation) to operate in an international environment, under conditions of a cross-border market and with an international professional orientation.
- It is strengthened primarily through inter-university agreements or agreements between bilateral or multilateral governments of exchange of different kinds.

7.6.2.5. The Strategy Enables

- Develop the option of missions and exchange of practices for students, teachers and staff.
- Active promotion of international training programs and scholarship opportunities.
- Establish financial aid to encourage academic mobility

7.6.3. Systemic Implications of Internationalization

- Subjects with international component: The traditional philosophical ideal prevails that knowledge must be transmitted beyond borders.
- Bilingualism-Multilingualism.
- Approval and standardization of Credit Systems.
- Curricular flexibility.
- Double Degree, Joint Programs.
- Foreign visiting professors.
- Internationalization activities at home.
- International Certification/Accreditation.

7.6.4. In Terms of Research

- International joint projects.
- International scientific events.
- International research networks.
- International publications.
- Scheme of international pairs.

7.6.4.1. In Terms of Management

- International Relations Office.
- Counseling Program.
- Internationalization Information System.
- Relationship policy (agreements).
- Membership to Associations and Intl Networks.

7.6.4.2. In Terms of Cooperation

Application to international financing Technical Cooperation and Technology Transfer.

International Trans frontier Cooperation.

Export of Educational Services.

Colombia needs to guarantee a growth rate in GDP, with a potential of 6 percent or more in a sustained and socially and environmentally sustainable manner. To achieve the above, therefore, it is necessary to advance in three fundamental dimensions:

- The innovation;
- The policy of competitiveness, productivity and improvement of Quality; and
- The revitalization of "locomotive" pulling sectors that, through their direct and indirect impact, lead growth and job creation.

7.7. Computational Models

7.7.1. The Visualization of the SP on Its Environments

It is presented in 2 views:

- The self-evaluation of all the tasks of the University, which we discussed in Chapter 6, related to University outputs with research projects and other dimensions of evaluations as effectivity, efficiency.
- the global model of relevance or significance that measures the implication of the lines of research with factors of Effectiveness and Convenience.

7.7.2. The Architecture of Pertinence

Below is presented the corresponding architecture.

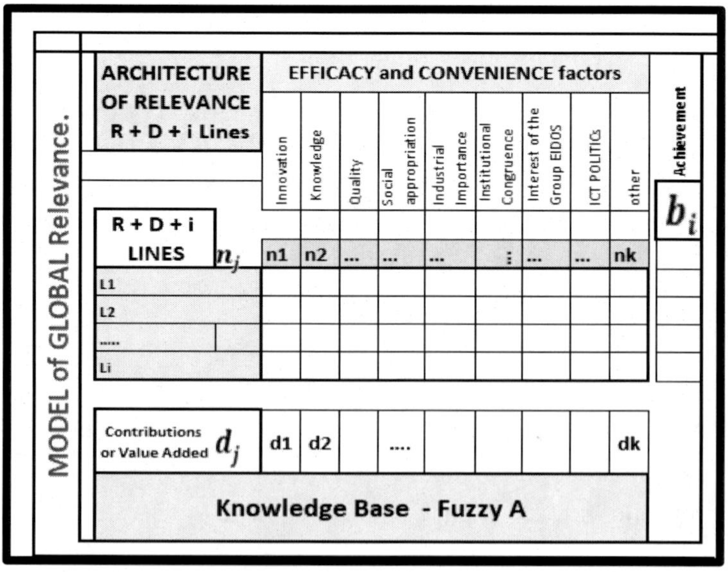

Figure 7.2. Architecture of pertinence.

7.7.3. The Computational Model

$\sum a_{i,j} \cdot n_j = b_i$ is the i^a achievement in terms of research and innovation of the Group's Production Portfolio, to obtain the fuzzy values $a_{i,j}$, regarding the opinion of the experts, which forms the Fuzzy Knowledge Base A. Where:

$i = 1, 2, 3$ ... Total lines of R + D + i

$j = 1, 2, \ldots k$. Efficiency and global convenience factors

n_j are the factors of fuzzy metrics.

$\sum a_{i,j} \cdot n_j = d_i$ representes the value of the contributions or added value of the j^a efficiency/relevance factor.

$i = 1, 2, 3$ ... Lines of investigation

$j = 1, 2, \ldots k$. Efficiency factors.

$a_{i,j} = b_i/d_j$ is the fuzzy technical coefficient: represents the amount of achievement of the I + D + I line per unit pf value added of the j^a of the factors of relevance, effectiveness and convenience.

Conclusion

A mathematical-computational model of external relations, or Social Projection of the FESSANJOSE has been presented in the space of the DIGITAL TRANSFORMATION, facing the IV Industrial Revolution, whose final purpose is a new model of University with its digitized components, in which the architecture enables to visualize its different layers: multiple dimensions, the knowledge, and the functionalities.

The achievements to be obtained with the general and specific SP architecture, as well as the methodology aspects were included.

In addition, the dimension of Internationalization, in its different aspects, was discussed.

RESOURCES

FESSANJOSE Documents of several curricular programs that have been qualified registered at MEN since the last 20 years. In each document, the whole 15 qualifications conditions are presented, among which one is related to the Social Projection and External Relations.

Pérez Gama, Alfonso & Quizza Tommich, Mandina. International Accreditation of the Engineering by Propaedeutic Cycles: An Imperative for Colombia. In: *EDUCON 2012 IEEE Global Engineering Conference,* Morocco, April 2012.

Pérez Gama, Alfonso (Editor). *"Knowledge Engineering: Principles, Methods and Applications"*. New York, London: Nova Science Publishers, 2015. ISBN-13: 978-1634639095. ISBN-10: 163463909X. Available in Amazon.com.

Pérez Gama, Alfonso. Analytical Models for Tertiary Education by Propaedeutic Cycles Applying Knowledge Engineering and Knowledge Management. In: Huei Tse Hou (Editor), *"New Research on Knowledge Management Models and Methods"*, chapter 2. Intech - Open Access Publisher, covering the fields of Science, Technology and Medicine. ISBN 979-953-307-226-4.

Chapter 8

CONCLUSION

8.1. EPILOGUE

8.1.1. Introduction

In this epilogue, a synthesis of U SAN JOSE-DIGITTRANSF is included, its origins in the FESSANJOSE, the description of the problems, the solution, the technologies used and the relevant aspects as well as the justification of the model, the implementation and related. Additionally, we would like to point out that on our website, www.ussanjose.edu.co, we present the demos of each of the DT tools, with their output Matrices of Results, with user manuals, management guides, and also a .ppt presentation. See Figures 5.8 and 5.9.

8.1.2. How Did Our Organization Work?

Before 2008 the U. San Jose, FESSANJOSÉ had no research group. As a T & T institution (Professional, Technical and Technological) it had been known among others for having obtained the best TP GRADUATE in Systems Engineering on 2 occasions, according to the ICFES. It had also obtained the Qualified Record (MEN) of all its academic programs. With the advent of the Propaedeutic Cycles Law 749 et seq.m with no national or international experiences known, not even the Ministry of Higher Education (MEN) had developed a propaedeutic example in this regard, FESSANJOSÉ attended the propaedeutic cycles in all its programs. T & T education appeared unpopular as the poor relative of Higher Education (HE), with little social credibility and no "right" to evolve. See Figures 5.14, 5.15.

8.1.3. What Were the Processes Carried out to Identify Our Situation?

1. The FESSANJOSÉ decided to create the San Jose EIDOS International Research Group to address the highly complex problems of post-secondary education in Colombia.
2. Also defined lines of research in Propaedeutics, Student Dropout and Permanence, Productivity, Competitiveness and High Quality Education.
3. The EIDOS Group was recognized and categorized by COLCIENCIAS.
4. In addition, this Group has participated with articles in world-class events (IEEE, ACM, ISTEC), and with national and international books and chapters, intellectual assets of software, and has participated in numerous national and international calls.
5. Up to day the Group have more than 80 products, in their majority certified by COLCIENCIAS, the National Directorate of Copyright, DNDA, and indexed in different databases such as Scopus, and IEEE Xplore, among others.
6. Continuous improvement of our DT architecture, including all the tools, methods and methodologies.

8.1.4. What Was Our Methodological Strategy?

It was based on the application of multiple paradigms with intensive use of high-level ICT with: Models (mathematical-computational), Systems and Software Engineering, Knowledge and Artificial Intelligence, including Multilevel Software Business Architectures Systems, SoS. We also start the development standards of the IEEE. Our first product was the Curriculum Coherence Matrix, in 2009, which we have evolved towards a very friendly and complex architecture, applying it to demonstrate and quantify the articulation of competencies and knowledge between propaedeutic cycles and to optimize the curriculum for the knowledge society to prepare the digital cyber-talent of the III Millennium: PT in Software Development (cyber-programmers), T (cyber-Technologists) in Software and Professional Engineering in Architectures Systems and Knowledge, (cyber-engineers). See Table 3.4.

8.1.5. What Was Our Target Scenario: Where Do We Want to Lead the Organization?

Towards a digitally transformed university, forming cyber-graduates at all levels facing the IV Industrial Revolution, pointing out a way to reduce both the digital and related digital talent gaps and the complexity of the Sector. See Figure 3.6 iMIS.

8.1.6. How Did We Plan to Achieve It?

With the digitalization starting from the Social Projection of the University, which is also one of the pillars of our DT model, with intelligent constructs, which has involved our in house developed Architecture of Dual Data-Knowledge Digital Transformation. See Figures 7.1, 7.2 and 3.6.

8.1.7. What Are the Main Triggers for Every Thought?

1. To build the university of the postmodernity at a higher level, that transcends the IV Industrial Revolution, with high quality; productive at its levels (organizational, educational and student), and competitive.
2. With the purpose of offering to Colombia and the world, a current, inclusive post-secondary education that confronts the socio-technical problems with exponential tools and which, according to the national Government, is aligned with the country's entry into the OECD, which implies, the use of best practices, transparency and social inclusion, among many others.
3. In addition, convince the MEN that you can skirt with leading instruments to reach excellence from fragile T & T education, although the MEN itself has been reluctant to these innovations.
4. In the Government of Colombia National Planning 2019-2023, one of the most important covenants for the technical, economics, industrial and social advancement, was registered as the Pact for Digital Transformation of companies, with an investment of 7 billion USD, which has encouraged us to continue improving continuously and decisively, implementing our DT models.

8.1.8. How to Know That We Arrive to the Desired Situation?

First of all by the knowledge, the control data of our Intelligent DT System (Figure 3.6), the cognitive metrics accompanied by numerous indicators and the High Performance Indicators (HPI) acceptability, with several awards, recognitions and distinctions, such as the medal awarded by the national Government through COLCIENCIAS, EMERITUS RESEARCHER, the Leadership Award of the IEEE Education Society, and also several awards at the level of our Seedlings of Research (national and international). See Figures 3.6 and 4.3.

8.1.9. When Has the Problem Been Eliminated?

The HE evolves permanently, especially since it must be nourished by science and technology: The sector does not remain solved, for which the social, economic, technical and political problems are permanently redefined in front of the obtained solutions that improve our performance; it also means that the environment changes and new situations appear that require new treatments and new strategies that complicate the HE. Our DT evolution and dynamic methodology responds to these new challenges.

8.1.10. And the Solution: How to Measure the Achievements?

With our Architectures, models and systems for the digitization of organizational and academic dimensions with enhanced implementation in situ:

1. Our DT dual model is supported by computational models that generate metrics of different order.
2. Our iMIS generates continuous management and operational control information.
3. All our constructs also generate cognitive metrics supported by fuzzy logic, in addition to different HPI.

8.1.11. What Was the Estimated Time of Our Project?

The DT is like a train journey: it has stations and stops, but not a final arrival as a traditional mono-project. The dynamics of the products and results of research gave us the pattern and support of the FESSANJOSÉ, represented in salaries, national and international travel support, the payment of participation rights to events, office supplies, equipment, communications, administrative support and related, estimated at $ USD 400’000,000. See Figure 3.10.

8.1.12. What Was the Methodology Used?

As we mentioned, the methodology is a research variable redefining the methods of:

1. Software engineering with evolutionary prototypes.
2. Knowledge Engineering and Artificial and Computational Intelligence.
3. Systems Engineering of the III Millennium and Data Science.
4. And also, our own complementary methodological constructs that are documented in our papers, books and book chapters.
5. Preparation of user manuals and management guides for our tools. See Figure 3.5 and 3.8.

8.1.13. How Did We Apply the Defined Methodologies?

1. Understanding and defining well the problems to solve, with their quantifications.
2. For each we developed plans to solve the problems that gave rise to the methodological construct itself.
3. External Consultant.

8.1.14. How Was the Solution Given to Each Problem?

1. "Making way to walk" as the Catalan singer-songwriter, used to say!
2. The Methodology: working it as a research variable.
3. Using architectures, mathematical and computational models that facilitate us the coherent processing with the quantification of important volumes of data and curricular and organizational knowledge.
4. The MEN in practice appears as the determiner of the curricula, and our claims have no answer: currently there is no third party to turn

to; MEN even promotes the disintegration of the HE with the Law of Tertiary Education, separating T & T in a way that is exclusive and despicable.

8.1.15. What Were the Main Challenges That We Faced?

1. Resistance to change and aversion to pedagogical innovations.
2. Insufficient budget.
3. Ignorance of cutting-edge technologies by academic peers and members of the CONACES Room of the MEN and the NCA.

8.1.16. The Result

Our model **USANJOSE-DIGITTRANF**, matured, documented, validated and implemented.

8.2. General Conclusions and Recommendations

8.2.1. General Conclusions

- The *membership of Colombia to the* OECD is a great opportunity for the DT in Colombia given the coherence of the OECD policies of productivity and competitiveness that are perfectly aligned with what the DT pursues.
- The good practices and the policies of the OECD lead us to a continuous and coherent improvement: Colombia will be evaluated with international standards and with proven and efficient review mechanisms.

- In this Post-Conflict process, the OED will help us to achieve a *peaceful* and *equitable country*.
- Also, help to reduce *social exclusion*, reduce the *digital divide* and *recommend* reintegrate the HE, that the tertiary Law of Tertiary Education broke apart.
- DT *technology* can be a disruptor or a facilitator. The power of the technology used in HE can be exploited not only through the implementation of new models and disruptive pedagogies, but also by innovating through the research, changing the way of doing things and adapting to the new educational and business environment.
- Many signals show that 1/3 part of investment returns is feasible in the whole world. The ICT digital challenge is the new business culture based on digital skills, competencies and expertise.
- In this book we have separated the organizational value chain into several critical elements that are at the center of the PSEd DT:
  1. The educational and business model;
  2. The institutional and academic structure;
  3. The teachers and the staff;
  4. The robotic academics and organizational processes;
  5. ICT capabilities with digitized social projection offer;
  6. And the model of commitment to society, government and industry.
- Therefore, understanding the need to transform learners to change society and also having a good understanding of what should change is important; but the key to success lies in the implementation of PSEd DT.
- Current young millennial users deserve special attention: it is necessary to listen to them, they who will see everything and educate using mobile phones with new pedagogic models. Nowadays, the social and economic sectors are being redefined in such a way that the daily life will change, including promoting modifications for a new politically, philosophically and occupationally world order. New skills which the IV Industrial Revolution demands are to be

included by default in the curricula of HE institutions in the short and medium terms, or such plans will become obsolete.

- DT under ICT focuses on business, prescriptive and prospective analytics (supported by *tools for future analysis*, TFA), agile and modern infrastructure, staging enough with modern capacity, improving the ample use of innovative IOT platforms, educational cloud apps, and many others.
- Using the *modern educational analytics*: How to generate greater value added through the data and knowledge, wherever possible, and at any opportunity? … and also including data, information and knowledge integration, unified data-types and objects, and involving different sources? Etc.
- DT makes intense use of management and performance metrics and indicators for follow purposes; e.g., applies ROI, conversion rates, qualified talent, condition of marginality, organic rate of repurchase (without investment in marketing) and SLA. Permanently evaluates the virtualized experience of the DT user.
- DT is not reengineering neither retro engineering: it is digital mapping, extended horizontally and vertically inside PSEd institutions. But how to transform is where most educational organizations do not succeed! While there is no single roadmap for all organizations in all sectors, we found that digital business agility capacity is related to positive financial performance as well as the ability to respond to digital disruption.
- The digital commercial agility is composed of several sub-capacities: (1) the ability to detect relevant trends and informed decision making based on evidence in a quick and collaborative way and fast execution; and (2) the ability to quickly translate decisions into action.

8.2.2. Recommendations

We present some concerns for institutions of different orders (*universities*, *industry*, *commerce* and the *government*): those that have matured in DT, the others that are beginning this long journey of DT, and those who are still thinking about it.

8.2.2.1. To the Universities

1. The HE will continue to make greater use of new technologies because this is where the greatest degree of transformation will occur to generate knowledge. In this sector, the opportunities are endless: as we mentioned several times, in the long process of the DT there is no line of arrival for publication of PSEd objectives to achieve cost savings or digitalize to improve the customer experience.
2. It is recommended to take advantage of the opportunity without prejudice to make the DT for PSEd, where there are endless possibilities to take advantage of automation platforms and ecosystems of users (people and companies) to bring to students and companies in this sector, that should be in an innovation constant.
3. What we are witnessing today is a seismic change in global educational models. Traditional programs are being completely interrupted. Visionary universities can see emerging opportunities and transform themselves to exploit these opportunities. Any large-scale transformation tends to alter and uproot existing structures and deep-rooted mentalities. This includes higher revenues to improve operational performance for successful expansion into new markets and customers. The digital technology itself, like any other business transformation, probably demonstrates the value that is expected to be generated over time.

4. It is very urgent the new definitions of modern educational model *pari passu* with pedagogic models which include analytics learning.
5. HE in developing countries, although is taking the first steps towards digital metamorphosis, probably has experienced a positive start in their digital journey, but sometimes with mixed results or results that are not on the expected scale.
6. It is recommended that PSEd institutions visualize technological change immediately to take position facing the IV Industrial Revolution.
7. The PSEd institution must take into account that probably the DT considering the impact of *the digital* in the different internal ecosystems of the *institution*. The above can be driven internally throughout the organization and externally through value chains for the end user.
8. Continuity plans at different levels must be developed that must include digital learning of the users' experience.
9. There is no single approach for DT in *universities*. There is a variety of architectures and operating models in the different pedagogical ecosystems based on the relevance for a given positioning, whether it is HQ or *global competitiveness*.
10. The digital-organizational learning and the DT addressing by research and innovation should be an imperative within the PSEd.
11. On the other hand, the long DT journey will probably come with challenges that may seem daunting, especially in the early stages of the DT, but with the enabling technologies highlighted above, the DT can be given a disciplined, systematic, academic and rigorous treatment, supported by solid national and foreign strategic alliances, and also by the management of the new digital talent facing the IV Industrial Revolution.
12. It is recommended that the PSEd consider that any large-scale transformation tends to alter and eradicate existing academic and scientific structure, academic programs, study plans and deep-rooted mentalities, which causes risks that must be taken into consideration.

8.2.2.2. To the Industry and the Commerce

1. In the new world of exponential technologies that are undoubtedly enablers of DT, some companies will survive and others will not. Companies that are agile, establish a culture of change and risk and also participate in a wide spectrum of customers, employees, markets and technologies, will prosper. Having a unique focus on growth is no longer a silver bullet, but rather the quality of relationships and the ability to participate commercially. This is the reality of a convergent world. DT is not a passing fad.
   - It is recommended that companies take advantage of DT's potential that is so overwhelming that there is no time for an industry to take the position of "wait to see what will happen." Digital skills are no longer the plus: they are the norm and there is need to develop them within the companies and with the support of the *university*.
2. Today's world exhibits a case of hyper-digitization of rapid development. Advanced data science, driven by high-impact intensive computer and cognitive systems such as machine learning, provides an unprecedented opportunity to unlock the value of interconnected data. Sensors in electrical networks and water systems, smart buildings and congested roads can optimize the use of scarce resources. Predictions based on information relayed from security cameras, satellites and ground can improve public health and safety.
   - It is recommended to the companies to invest in the new oil *the data*.
3. Among companies that take a proactive position in the digital revolution and take advantage of the full potential of disruptive technologies are those who want to:
   a. Reform the customer's value propositions.

b. Reshape their operations with robotization of the academic and institutional processes to deliver new value propositions to the client in an effective, efficient and innovative way.
c. Or do both at the same time, which leads to the broader transformation of the industry.

4. Those who do not take advantage of the new digital era can drastically limit opportunities for better future achievements in terms of competitive and income. As industry leaders, they have the opportunity to distance themselves from new and existing competitors.
   - It is recommended that companies be proactive in the DT.
5. Taking into account the digital divide in digital talent at all levels, which requires the successful entry into the IV Industrial Revolution:
   - It is necessary to design contingency plans with agreements between PSEd, industry, commerce and the *government* to train digital talent with the new competencies demanded.

8.2.2.3. To the Government

1. Some official institutions will naturally focus more on innovation and others will focus on the steady state. The more interconnectivity they have as a group in these different priority areas, the stronger they are as an ICT organization, and the stronger they will be as collaborators within the business in general.
   - It is recommendable for the government the idea of developing the ICT DT organization as a whole, instead of creating a DT innovation group; together, with a stable state group, should be better.
2. The economic and technical efforts made by academic institutions and companies are very important, therefore, it is required to:
   - Create and carry out economic and fiscal incentives to favor the development of the DT in the productive sector and other social sectors, towards the competitiveness of the State. Even more so,

for those PSEd that require to finance at least partially the expensive DT of quality. The international consultancy should be supported without hesitation to internalize the experience of advanced countries in DT.

Chapter 9

APPENDIX

9.1. DT TECHNOLECTUM

Analytics learning: deals with the application of machine learning for the discovery, interpretation and communication of pedagogical patterns and significant learning in the data. It is especially valuable in areas rich in recorded information, where the analysis is based on the simultaneous application of statistics, computer programming and operations research to quantify the apprentice's performance.

Analytics: is the use of quantitative methods to process data to reveal patterns, create new data, information and knowledge, and support mainly decision making for DT *governability*. Historically the term "analytics" is referred to processing structured data using solutions like *operations research* and business intelligence (BI) to provide such discernments.

Architecture: architecting implies the organization of subsystems or components to obtain better and/or new functionalities. If any component is knowledge-based, an intelligent behavior is obtained. The architecture multilayer approach is a system of systems (SoS) one, that ensures compliance with government policies, rules and standards, in a highly complex social institution with intellectual assets and knowledge processes, which is a usual situation in a university institution.

Cloud computing: is a type of Internet-based cognimatics that provides shared computer processing resources and data in cloud to computers and other devices on demand.

Collaboration: functioning as a team to accomplish some process such as design or engineering change approval.

Concurrent engineering (CE): for DT is the systems of services and the culture necessary to support concurrent engineering. The four CE environment factors are: organization, communication infrastructure, requirements, and product development.

Configuration: the architectural grouping of components (subsystems) or outcomes as defined by the number, nature, and interconnections of its constituent parts for getting functionalities.

Content management: the management and control of all forms of electronic data, knowledge and information. Content management with security includes data storage, access control, and some level of version control.

Digital mapping: is the process by which a collection of data of subsystems are assembled and formatted into a digital construct, i.e., Digitization.

Digital transformation (DT): is a complex process that takes place in long days, which is required for the successful survival of an organization to the IV Industrial Revolution, with the final purpose of being very competitive, productive and of high quality. It is the way to the 4th Industrial Revolution.

Digital twin: a physics-based description of a system resulting from the management and application of data, models, knowledge and information from confident sources across the system's lifecycle.

Digitization: the process of moving information from "paper" to digital forms (e.g., computer files, connected data elements representing a product, managing a configuration and all its dependencies) to transform business, moving to a digital business. Also, virtualization of services and real processes using, e.g., expert systems.

Disruption: the digital interruption, which is driving the transcendental transformation in the PSEd that responds to the risks and threats of digital interruption.

Document management: the data architecture processes, procedures, and tools focused on creating, maintaining, and managing documents. Comprehensive applications are available that provide authoring, scanning, structuring, rendering, and printing of documents.

Ecosystem: is defined inside PSEd as a network of HEI agents, accredited or not accredited, that are articulated around the DT from the needs of the country that respond to the PSEd contexts' characteristics in order to increase the student, educational, academic and institutional productivities to finally make the transfer of knowledge in function of the substantial improvement of the HE competitiveness that integrates all the PSEd's levels without exclusions, for the social and productive development of the country.

EIDOS: research group originated in the National University of Colombia, Dept. of Systems, in August 1988 and finalized with a research project presented and approved at COLCIENCIAS and executed by the Group. In 2008 some EIDOS teachers linked to FESSANJOSE and created the *San José EIDOS International Group* at the beginning of 2008 as an extension of EIDOS; the new Group is recognized and categorized by COLCIENCIAS 0084828. The Group has more than 80 publications and digital constructs. Our allied partners are the international consulting companies BI Planning and APHROS.

Enterprise resource planning (ERP): systems that help organizations plan their manufacturing processes and control

Enterprise: a portion of a company that is related by a common interest in a product or group of products. An enterprise may also logically include a network of subcontractors involved in the common product. PDM systems are often applied across an enterprise.

Excellence: is about doing your best results. In addition, every day, we are reminded of how important it is to strive for excellence, be it in life or at work. By nurturing a culture of excellence within your organizations, you open the path to success.

Federated data architecture (FDA): a form of system structure that accommodates multiple heterogeneous databases. FDA recognizes that information required to execute business processes usually resides in multiple repositories, which often represent a completely heterogeneous mix of technologies and implementation approaches. *Academy and institutional.*

FESSANJOSE: (Fundación de Educación Superior San José) university at Bogotá which is a T/T institution redefined by *propaedeutic cycles.*

Framework: a common user environment for cooperative use of tools to support a type of user. The environment is built from a collection of software program facilities and a set of extensible services.

Graphical user interface (GUI): a user interface to a software that includes windows, menus, dialog boxes, icon palettes, and other point-and-select command selection methods, and the use of graphics to present and manipulate information.

Higher Education (HE): it is post-secondary or undergraduate education, which includes the levels: *technical-professional*, *technological* and *professional.* Then following them, there are the postgraduate levels: Masters and PhDs.

Integration: test method for interconnecting application software. One set of customer interface menus are used to perform common tasks in both programs so that the customer does not have the insight that they are leaving their tool. All types of data in the data architecture are shared including metadata, objects, documents, product structures, etc.

IV Industrial Revolution: it is the new era that is characterized by the convergence of several technologies from the real world with the virtual world, where the boundaries between ICT, physics and biology are diffuse.

Lifecycle: the description of the distinct phases through which each product passes during its product life. This includes stages such as requirements definition, concept design, production, etc.

Machine learning: is an AI field for automatic learning, which uses algorithms and statistical techniques to obtain computer system constructs with the ability to "learn" from the data, without being explicitly programmed.

Metadata: information about the data controlled by the PDM system. For example, digital drawing number is metadata about a digital drawing.

Methodology: refers to a systematic process and methods of reasoning and thinking in order to express and represent a problem and obtain an appropriate solution. It also refers to the way of planning and determining problems: it is the art of problemtization. It is also the way to build a solution and, at the same time, to verify that the solution is correct. Also, *a methodology* is defined as a collection of related processes, plans, methods and tools used to define and solve problems.

Metric: a quantitative measurement of a system, component or process to determine the degree to which it possesses a given attribute. Metrics are often used to measure quality and process improvement.

Model-based engineering (MBE): is an approach to engineering that uses mathematical models (not only CAD models) as an integral part of the technical baseline definition that includes the whole lifecycle.

Post-secondary education (PSEd): see HE.

Process: (noun) an ordered sequence of steps performed for a given purpose, such as the software development; **(verb)** to perform an operation on data, knowledge and information.

Product data management (PDM): solutions and methodologies used within an enterprise to organize, access and control data related to its products, and to manage the lifecycle of those products.

Product: a generic term for whatever is produced by a process, whether goods or services. Any deliverable item of a project or program.

Program management: the planning, directing, and controlling of human and non-human resources to achieve specific program or project objectives. Also known as *project management.*

Program portfolio management: focuses on managing the group of projects (a program) that a company executes to create, build, and support their product suite.

Quality assurance: a planned and systematic approach to determine that a product conforms to established technical requirements.

Quality: the degree to which a system, component, or process meets specified requirements or meets customer expectations. It is the difference between excellence and the actual level of a system performance.

Reporting and analytics: the process and tools that support analyzing and searching through data, and producing reports based on the results.

Requirement, functional, logical and physical (RFLP): an acronym from the *systems engineering* community INCOSE (www.incose.org), that refers to defining requirements, and then mapping those requirements to the functional, logical, and physical instantiations of the system.

Requirements management: gathering and managing user, business, technical, functional and non-functional requirements within a product development project.

Return on investment (ROI): the value of an investment's profit to the organization, an indicator expressed as a percentage.

Revision level: the current authorized and documented level of a component, configuration, or text. It is defined and updated according to configuration management change.

Revision: a modification of an item during the process of developing a product. Many organizations use some form of code in the item number to indicate which revision of the item is being examined. Item revisions may also produce a variant item, that is one that is different for some specified reason.

Road map: action plan including dates.

Role: describes responsibilities associated with a specific job type. Examples include teacher, project manager, designer, and reviewer.

Scenario: set of circumstances that are considered the environment of an organization, person or event (i.e., locus). It is also associated with the site where a DT plan is developed.

Self-assessment: self-analysis of the performance results of an organization.

Social Projection: The social projection of the university is one of the fundamental pillars of our DT model. It requires the mapping of products, services, knowledge, consultancies and so on, on society, industry, government and even the rest of the world. In synthesis it refers to external relations, which obviously includes internationalization

Software as a service (SaaS): a means of software delivery in which all functionality is provided over the Internet or *virtual private network*.

Strategy map: methods and tools that provide a macro and simplified view of an organization's strategy and provide a language for communication to describe it. They constitute a graphic summary of the strategy of both the corporate and the strategy of the DT.

System: is a combination of interacting elements organized to achieve one or more stated purposes. It may also be an integrated set of elements, subsystems or assemblies that accomplish a defined objective. These elements include products (hardware, software, firmware), processes, people, information, techniques, facilities, services, and other support elements.

Tertiary education: post-secondary education, including the levels of *professional technician*, *technology* and *professional*. See HE.

Top-down design: a design method that combines the use of divide-and-conquer techniques with one of adding increasing detail, as the design processes. Top-down is typically used at the beginning of the design process.

Version: a variation of an approved and controlled item. Multiple versions of the same *revision level* may be created. Versions allow variations of components, configurations, and documents to be developed and used without requiring formal changes to the *revision* or *baseline* of an item.

Workgroup: any group of people working toward a common goal as a team.

9.2. KPI: Key Performance Indicators

In the domain of the social sciences (HE), in the digital transformation predominate the flexible systems that can adopt different states without changing their identity. The measurement and quantification represent very important challenges between attributes of quality and qualitative statements. The measurement in the processes dimension necessarily precedes the explanation of the performance of educational and institutional variables and of the same system, and in a special way the prediction and prospecting of what concerns us: control of the desertion, and optimization of the student permanence towards the productivity and the high quality.

The monitoring of academic and administrative management through indicators, presents comparative advantages over other systems. The synthesis, the absolute and relative valuation, the precision and other no less relevant aspects, mark the difference in the presentation, the follow-up to the evolution, the evaluation and judgment of the results, and also, the monitoring of the decisions that must be reflected in the indicators.

It is then necessary, as far as possible, to bring indicators to management, carried out by the computer and cognitive component of the institution, with emphasis on the activity of the *systems unit* in the administration, support and securities of the infrastructure and information systems.

9.2.1. Indicators of KPI Management and Index Numbers

KPI are qualitative or quantitative metrics that help us identify the performance of a certain activity, process, strategy, and others. These metrics indicate our level of performance based on the objectives, purposes and guidelines that we have set previously, according to their frequency, category and nature, among others. An *index number* is a statistical metrics of changes in a representative group of individual data points. These data may be derived from any number of sources including institution and individual performance, prices, productivity, management and employment, compared with base value. An *indicator* is a tool used in the technical of socio-economics analytics in institutions, and, a concept in statistics, research design and social sciences, as education.

Institutional control (IC) is a clear contribution to the performance of the company, which depends on the degree and quality of the applied control and self-control exercised, the function, the management of the processes, the management indicators, and the productivity. Analytically, it is described as shown in Figure 9.1.

IC = *f*(CD, t (r), ACD, t (r), Ind(g), Prod, ...)

Given:

CD, t (r) = *f*(RM, (1/rms (detected errors)):
The degree and quality of the controls on the *resources* (**r**) in the *career directions* (**CD**), in a period of time (**t**), directly proportional to the mitigated risks (**RM**) and inversely proportional to the mean square errors detected in aggregate form.

ACD, t (r) = Existence of Self-regulation or Self-control of the author.
It means the degree and quality of self-control of the active students of an academic program (**D**), in a period of time **t,** expressed in the form of an aggregate index number.

On the level of *university welfare*:
Prod = *f*(x/r) Productivity is directly proportional to the academic/institutional production and inversely to the resources used.
Ind(g): the management indicators (KPI) as an aggregate function.

Figure 9.1. KPI conceptual framework.

The general indicators are data from *external* sources, or exogenous to FESSANJOSE that may or may not be related to its PEI, missional and complementary objectives. Strictly speaking, they cannot be considered indicators but as general statistics, opinion polls, etc. The perception of parents about the support services provided on a university campus is a general indicator because, although it is an evaluation, it is an external opinion and may not relate *per se* with institutional objectives. Analytically, it can be expressed in the plan of the Institution as follows:

KPI (performance or efficiency) are critical measures that require a benchmark, a standard or an objective against which to compare performance. They imply a relative comparison between data. An example of a performance indicator is that which results from the comparison between the number of *graduates of a program/career* and the number of total students (enrolled) in a given semester.

The references could be in terms of:

1. *Inputs*, *raw materials*, *ingredients* or elements that enter the system, such as human resources (teachers, administrative staff), financial and physical resources that enter programs, activities or services of a university. The percentage of ICT resources that a Faculty allocates to teaching and learning is an input indicator.
2. *Processes*, which are the activities, forms and methods used to execute programs, activities and services and that have a quantifiable product. The process indicators inform about the way in which the inputs become products or results with their metrics. Among others, they are indicators of processes: student performance evaluations, used technologies, measurements of student performance (e.g., self-control board), teacher compliance, quality assessment of teaching, student tracking systems (student model), etc.
3. *Results*, ex-sums, outputs, products, effects or impacts that emerge from the system. The following are indicators of results: the index of the number of graduates per program in period t, the number of degrees awarded in a given semester/year, the impacts that occur in the community because of the qualified registration granted by the national or international accreditation program.

9.2.2. Desirable Features

Indicators and indices of KPI should reflect the measurement of educational quality to reflect the cause-effect relationship between spending and products. So, we could establish these characteristics:

- Relevance: must be useful for the governance of decision making.
- Verifiability: the interpretation given to the indicator must be uniform for its *users*. It must be traceable to check its objective.

Socialization is important within the entire academic community and *staff*.

- No bias: the calculation of the indicator should be free from personal or statistical bias.
- Quantifiable: must have *metric* and *cogni-metrics*, to be estimated by numerical or qualitative quantities as a result of the measurement and or qualification.
- Visibility: they should be published in the *administrative portal*.

9.2.3. Analytics of Digitized Institutional and Academic Processes

The KPI taxonomy is done according to:

- Objectives of the company, departments and people including students, which requires strategic alignment with the *corporate plan* and also take into account managerial and business styles.
- Orientation: towards academic and administrative *management* or towards administrative and operational control.
- Frequency of *open* KPI: i.e., hourly, daily, weekly ... *closed* KPI from month to day ...
- Variability: *static* KPI (states, levels) and *dynamic* KPIs (rates, %, etc.).
- KPI categories, i.e., financial, safety, qualitative, quantitative ...
- Nature of the KPI: *quantitative* or *qualitative*; *prospective*, *proactive* and *reactive*.
- The KPI *purposes*.
- Understanding performance so that KPI guide us to success.
- Identify the areas or departments susceptible to improvement according to the mission of the organization.
- Achieve indications of efficiency and effectiveness to achieve the goals of economic and academic use of the institution.

- Set levels of personal, departmental and *institutional* performance that mark the strategic objectives.

From the above it is important to highlight:

- The *internal* KPI are those that reflect the inputs within the *institution*. By way of example:
  - Percentage of students by program, area and by subject.
  - Graduation rate per propedeutical cycle.
  - Number of financial aids or subsidies requested per student and/or financial institution.
  - Coverage rate of students benefited by program/cycle.
  - Repetition rate per subject per program/cycle.
  - Attrition rate per subject per program/cycle.
- The *external* KPI reflect the situation of the *institution* in the national/international context:
  - Acceptance rate of graduates in the labor field by program, propaedeutic cycle or level.
  - Research groups registered/recognized by COLCIENCIAS.
  - No. of publications and appointments per teacher.
  - No. of intellectual property registries, intellectual software assets, patents, inventions, ISBN, ISSN, etc.
  - Consultancies contracted by program and faculty.
  - No. of memberships to scientific and professional societies, awards, membership in specialized societies.
  - No. of presentations at local and international conferences.
- The *operative* KPI include information and proportions that reflect the operation of the *institution* and its productivity:
  - Unit costs per program/propaedeutic cycle: active-deserter students.
  - Number of teacher/students per program/propaedeutic cycle.
  - Number of students per subject per program/propaedeutic cycle.

- Available courses in person/virtual modes by program/ propaedeutic cycle.
- Availability of bibliographic material: volumes per career, journals by program/propaedeutic cycle.
- Rates of use of the library and of bibliographic databases per month and per program/propaedeutic cycle.
- Availability and use of laboratories by program/propaedeutic cycle.
- No. equipment/jobs per laboratory per program/propaedeutic cycle.
- No. diplomas/seminars and courses for graduates/not graduates by program/propaedeutic cycle.

9.3. Acronyms

AI:	artificial intelligence.
ANDICOM:	ICT International Congress, Cartagena-Colombia.
BI:	business intelligence.
CINTEL:	International Center FOR Research In Telecommunications (located in Colombia)
COLCIENCIAS:	Colombian Administrative Department of Science, Technology and Innovation, www.colciencias.gov.co.
CONACES:	National Intersectional Commission for Quality Assurance in Higher Education, Colombia.
CONPES:	National Council of Economic and Social Policy, Colombia.
DE:	digital engineering.
DNDA:	National Directorate of Copyright, Colombia.
DT:	digital transformation.
EA:	enterprise architecture.
EMM:	end-means matrix.

ES:	expert systems.
EY:	Ernst and Young, www.ey.com.
FDA:	federated data architecture.
FDM:	fuzzy dashboard matrix.
FESSANJOSE:	Fundación de Educación Superior San José, U. San José, Bogotá, http://usanjose.edu.co/.
FL:	fuzzy logic.
GUI:	graphical user interface.
HE:	higher education.
HEI:	HE institutions.
HPI:	high performance indicators
HQ:	high quality.
ICT:	convergence of information technologies (cognimatics, computing and telecommunications, particularly the Internet and computer networks) with their technological base and information or "content".
IE:	inference engine.
IEEE:	institute of Electrical and Electronics Engineers.
IEP:	institutional educational project.
IK:	intellectual capital.
IKBS:	intelligent knowledge base systems.
iMIS:	intelligent management information system.
IOT:	Internet of everything.
IS:	intelligent systems.
KB:	knowledge base.
KE:	knowledge engineering.
KEE:	KE in education.
KPI:	key performance indicators.
KS:	Knowledge Society.
MBE:	model-based engineering.
ME:	middle education.
MEN:	Ministry of National Education (Ministerio de Educación Nacional), Colombia, www.mineducacion.gov.co.

MinTIC:	Ministry of Information and Communication Technologies (Ministerio de Tecnologías de la Información y las Comunicaciones), Colombia, www.mintic.gov.co.
MIR Matrix:	matrix of impacts, indicators and results.
NCA:	National Council for HE Accreditation (Consejo Nacional de Acreditación), Colombia.
O/L:	on line.
OC:	organizational culture.
OECD:	Organization for Economic Co-Operation and Development.
PDM:	product data management.
PSEd:	postsecondary education.
QA:	quality accreditation.
RD&I:	research development plus innovation.
RFLP:	requirement, functional, logical, and physical.
ROI:	return on investments.
SaaS:	software as a service.
SLA:	level of service.
SoS:	system of systems.
T&T:	(T/T) technical and technological levels of HE.
TFA:	tools for future analysis.
U:	university.
UI:	user interface.
VVTL:	validation, verification, testing and legitimization.
WEF:	World Economic Forum, www.weforum.org.

References

Arias B., Y., Rosete S., A., and Martínez R., R. (2006). *Propuesta informática para seleccionar personal por competencias utilizando técnicas de inteligencia artificial.* [*Computer proposal to select personnel by competences using artificial intelligence techniques*]. In *Ingeniería Industrial*, 27 (2-3): 33-39. La Habana, Cuba: Instituto Superior Politécnico José Antonio Echeverría. ISSN: 0258-5960. [Online] https://bit.ly/2JpMqpf.

Berman, S. J. and Bell, R. (2011). *Digital transformation: Creating new business models where digital meets physical* (Executive Report). Somers, NY, U.S.A.: IBM Global Business Services - IBM Institute for Business Value. [Online] https://goo.gl/hb8Ui7.

Blask, C. (2017). *Capitalizing in digital transformation in industrial enterprises.* Cartagena-Colombia: ANDICOM.

Cámara de Comercio de Bogotá – Foro de Presidentes. (2018). *Hacia una economía tecnológicamente avanzada y sostenible: Partitura para el desarrollo productivo de Colombia/Think Tank Transformando a Colombia* [*Towards a technologically advanced and sustainable economy: Score for the productive development of Colombia./Think Tank Transforming Colombia*]. Bogotá, Colombia: Cámara de Comercio de Bogotá. URI: http://hdl.handle.net/11520/19931.

Casale, F. (2017). *Colombia en las cadenas globales de valor.* [*Colombia in the productive value chains*]. Summit of Digital Trasnformation (Irpa Ai, Weston Jones Ey). Medellín-Colombia: ANDI, May 24.

CONEAU - Comisión Nacional de Evaluación y Acreditación Universitaria (2016). *Evaluación institucional: Criterios y procedimientos para la evaluación externa.* [*Institutional evaluation: Criteria and procedures for external evaluation*]. 1a ed. ampliada. Buenos Aires: CONEAU. ISBN 978-987-3765-15-5. [Online] https://bit.ly/2y9sP7J.

Cuyamaca College (2013). *Institutional self-evaluation report of educational quality and institutional effectiveness in support of reaffirmation of accreditation.* El Cajón, CA, USA: Cuyamaca College. [Online] https://bit.ly/2Miew3k.

Davis, G. and Olhson, M. (1987). *Sistemas de información gerencial.* [*Management information systems: conceptual foundations*]. 2nd. edition. USA: McGraw Hill LatinoAmericana. Translation from English to Spanish by Alfonso Pérez Gama.

Digital Transformation Initiative, DTI Project. (2016). *Geneva: World Economic Forum (WEF)/Accenture.* [Online] http://reports.weforum.org/digital-transformation/.

Espinoza D., O, and González F., L. E, et al. (1994). *Manual autoevaluación para instituciones de educación superior: Pautas y procedimientos.* [*Self-evaluation manual for higher education institutions: Guidelines and procedures*]. Santiago, Chile: Centro Interuniversitario de Desarrollo (CINDA), OEA. Colección Gestión Universitaria, ISBN 956-7106-19-3. [Online] https://bit.ly/2LAcj24.

European Association for Quality Assurance in Higher Education (ENQA), et al. (2015). *Standards and guidelines for quality assurance in the European higher education area (ESG).* Brussels, Belgium: EURASHE. ISBN 952-5539-04-0 (paperbound). ISBN 952-5539-05-9 [Online] (pdf) http://www.enqa.eu/index.php/home/esg/.

Eurydice European Unit. (2008). *Higher education governance in Europe: Policies, structures, funding and academic staff.* Brussels, Belgium: Eurydice European Unit. ISBN 978-92-79-08524-6. DOI 10.2766/29900. [Online] https://bit.ly/2y03Fbm.

Eyring, M. J., Joknson, M. W. and Nair, H. (2014). New business models in emerging markets. In *IEEE Engineering Management Review*, 42 (2): 19-26, Second Quarter, June. ISSN 00360-8581.

Font, R. (2017). *Transformación empresarial, más allá de lo digital.* [*Business transformation, beyond the digital*]. In *Foro 3C*. Bogotá-Colombia: CINTEL.

Gill, M. and Van Boskirk, S. (2016). *The digital maturity model 4.0 benchmarks: Digital business transformation playbook* (Report) USA: Forrester, Jan. 22. https://go.forrester.com/.

Hafsi, M., Assar, S. (2016). What enterprise architecture can bring for digital transformation? An exploratory study. In *IEEE 18th Conference on Business Informatics*. IEEE: Paris, France. ISSN: 2378-1971. DOI: 10.1109/CBI.2016.55.

Kukita, S. (2017). *Democratization of digital transformation.* NEC. Cartagena-Colombia: ANDICOM.

Legua García, L. (2011). *Acreditación y Centros de Excelencia* [*Accreditation and centers of excellence*]. Perú: Ministerio de Salud. [Online] https://bit.ly/2MeltT2.

Marra, M., Di Biccari, C., Lazoi, M. and Corallo, A. (2018). A gap analysis methodology for product lifecycle management assessment. In *IEEE Transaction on Engineering Management*, 65 (1): 156-167. Feb. ISSN 0018-9381. DOI: 10.1109/TEM.2017.2762401.

Martínez N., M. et al. (2016). *Primer manifiesto transformación digital.* [*First manifesto digital transformation*]. ANDICOM 2016 ICT International Congress. Cartagena-Colombia: CINTEL-Accenture.

Mejía, L. F. (n. d.). *Esquema de financiación de la economía digital.* [*Financing scheme of the digital economy*]. Bogotá, Colombia: Departamento Nacional de Planeación.

Moavenzadeh, J. (2015). The 4th Industrial Revolution: Reshaping the future of production. In *DHL Global Engineering Manufacturing Summit*. Amsterdam: World Economic Forum, Oct 7.

Morin, E. (1999). *Seven complex lessons in education for the future.* Paris: UNESCO. [Online] http://unesdoc.unesco.org/images/0011/001177/117740eo.pdf.

Nilsson, N. J. (2010). *The quest for artificial intelligence: A history of ideas and achievements.* UK: Cambridge University Press. ISBN: 9780521122931. Web Version, Stanford University: https://ai.stanford.edu/~nilsson/QAI/qai.pdf.

Nylen D., and Holstrom J. (2017). Digital innovation strategy: A framework for diagnosing and improving digital product and service innovation. In *IEEE Engineering Management Review*, 45 (3): 128-138. Third Quarter, sep. ISSN 0360-8581.

Olier, J. F. and Costa, R. (2018). *Nivel digital, directivos: habilidades gerenciales en Colombia (Resumen ejecutivo).* [*Digital executive level: management skills in Colombia (Executive summary)*]. Barcelona-Spain: EADA Business School. http://www.ey.com/.

Oliván Cortés, R. La Cuarta Revolución Industrial, un relato desde el materialismo cultural [The fourth cultural revolution, a tale through cultural materialism]. In *URBS. Revista de Estudios Urbanos y Ciencias Sociales*, 6 (2): 101-111. Zaragoza, España. ISSN 2014-2714. [Online] https://bit.ly/2kY0BDg.

Park, J., Lee, J., Lee, O. D., and Koo, Y. (2017). Alignment between internal and external IT governance and its effects on distinctive firms performance: An extended resource-based view. In *IEEE Transactions on Engineering Management*, 64 (3): 351-364, aug. DOI: 10.1109/TEM.2017.2718798. ISSN 0018-9391.

Park, W. Y., Ro, Y. K. and Lee, Y. H. (2018). How much does design internationalization matter in the face of technological change? In *IEEE Transactions on Engineering Management* 65 (2): 264-275, April. DOI: 10.1109/TEM.2017.2690320. ISSN 0018-9391.

Perdomo, C. (2017). *Transformación digital en retail: 5 estrategias para generar el cambio.* [*Digital transformation in retail: 5 strategies to generate change*]. Cartagena-Colombia: ANDICOM.

Pérez G., A. (1989). *Diseño de sistemas expertos en educación* [*Design of expert systems in education*]. Bogotá: U. Nacional - Facultad de Ingeniería - Departamento de Ingeniería de Sistemas.

Pérez G., A. (Ed.) (2014). *Knowledge Engineering: Principles, Methods and Applications.* New York: Nova. ISBN-13: 978-1634639095; ISBN-10: 163463909X.

Pérez G., A. and Joyanes A., L. (Eds.) (2001). *Sistemas de información e ingeniería de software en la sociedad del conocimiento.* [*Information systems and software engineering in the knowledge society*]. Bogotá. ISBN 958-97030-0-3.

Pérez G., A. and Medina, V. H. (1998). *Diseño de sistemas y prototipos con el apoyo del computador.* [*Design of systems and prototypes with the support of the computer*]. 2nd edición. Bogotá: U. Nacional - Facultad de Ingeniería/Asociación Colombiana de Informática (ACCIO). (Libro electrónico).

Rivera C., A. (2014). *Modelo para determinar el estado de un sistema de información de tipo académico: un análisis multicriterio.* [*Model for determining the state of an academic information system: A multicriteria analysis*]. Tesis de maestría. Bogotá-Colombia: Universidad Nacional de Colombia, Facultad de Ingeniería, Departamento de Sistemas e Industrial. [Online] http://bdigital.unal.edu.co/46219/.

Rolston, D. (1991). *Principios de inteligencia artificial y desarrollo de sistemas expertos.* [*Principles of artificial intelligence and expert systems development*]. USA: McGraw Hill Latino Americana. Translation from English to Spanish by Alfonso Pérez Gama.

Shaw, A. (2004). *University research centers of excellence for homeland security: A summary report of a workshop.* Washington, DC., USA: The National Academies Press. ISBN 0-309-09236-4. https://doi.org/10.17226/10984.

Sitoh, M. K., Pan, S. L. and Yu, C. Y. (2014). Business models and tactics in new product creation: the interplay of effectuation and causation processes. In *IEEE Transactions on Engineering Management*, 61 (2): 213-224, May. DOI: 10.1109/TEM.2013.2293731. ISSN 0018-9391.

Slotnisky, D. J. (2016). *Transformación digital: cómo las personas y las empresas deben adaptarse a esta revolución.* [*Digital transformation: how people and enterprises should adapt to this revolution*]. Buenos

Aires-Argentina: Digital Houses - Coding School. ISBN 978 987 502 400-7.

Zimmermann, A., Jugel, D., et al. (2016). Architectural decision management for digital transformation of products and services. In *Complex Systems Informatics and Modeling Quarterly*, 6: 31-53. March/April. RTU Press. ISSN: 2255-9922. [Online] https://csimq-journals.rtu.lv, http://dx.doi.org/10.7250/csimq.2016-6.03.

About the Authors

Jesús Alfonso Pérez Gama

MSc (Artificial Intelligence) U. OF ESSEX, UK.; M. Systems Eng. at the U. Nacional De Colombia Bogota; E.E. U. Distrital Francisco Jose De Caldas Bogota. IEEE Institute of Electrical and Electronics Engineers Life Senior Member; Member of the NY Academy of Sciences; Member of the International Input Output Association; Member of the International Council on Systems Engineering; Chair Asociación Colombiana de Informática ACCIO. Member of Colombia Systems Engineering Association ACIS. Emeritus Researcher Medal (Gov. of Colombia-COLCIENCIAS). Universidad Nacional De Colombia: Tenure Professor and University Master Fellow (Retired). Currently: Fundación de Educación Superior San José Chair of San José EIDOS Research Team.

Luis Giovanni Rozo Pardo

Psychologist of the U. Nacional of Colombia, Edumatics Specialist, and Database Management Technological Specialist. Researcher-teacher at San José Higher Education Foundation and co-director of the San José EIDOS International Research Group, participating mainly in projects about DT intelligent curricular and educational institutions management tools, and several publications and software developments. With interdisciplinary research and job interests into the field of combining aspects of human and natural sciences and technologies to use them for better tools in job and human development and cognition. And with professional experience in academic and technical writing and presentations, user-focused software prototyping, and cognitive hypermedia documentation tools.

Luis Carlos Gutiérrez Martínez

Doctor Honoris Causa in Education, Univ. Cundinamarca; MSs. in Innovation; Systems Engineer; around 20 years of experience in the educational sector; with specific skills in management and educational management, planning, monitoring and control of projects with emphasis on administration and curriculum management, cost and market analysis. Experience in the control and implementation of budgets of higher educational institutions in both: public and private sectors; entrepreneur in high performance companies in the field of R+ D + i projects. Teaching experience of fifteen (15) years, researcher in topics related to Big Data, curriculum, technologies in education, expert systems in education and cognitive analysis.

INDEX

C

D

E

K

L

M

R

S

T

U